G000320507

CATS

A Guide to Breeding and Showing

ANGELA SAYER

CATS

A Guide to Breeding and Showing

ANGELA SAYER

B.T. Batsford Ltd. London

© Angela Sayer 1983
First published 1983

All rights reserved. No part of this publication
may be reproduced, in any form or by any means,
without permission from the Publisher

ISBN 0 7134 4051 1

Filmset by Servis Filmsetting Ltd, Manchester
and printed in Great Britain by
The Pitman Press Ltd, Bath
for the publishers
B.T. Batsford Ltd
4 Fitzhardinge Street
London W1H 0AH

Contents

Introduction

EVOLUTION

Every member of the cat family, including the domestic cat (*Felis catus*), traces its ancestry back to a weasel-like creature named *Miacis*, which evolved some 40 million years ago in the Eocene period. *Miacis* was also the forebear of several other carnivorous groups including bears, civets, dogs, hyaenas, mongooses, raccoons and weasels. The domestic cat seems to have descended from *Miacis* through the civet, and while one branch was maintained, giving us the modern civet family, the offshoot developed and diversified. In evolutionary terms, the change from *Miacis* to cat was very abrupt, taking a mere million or so years, whereas ten million years were needed for the debut of the dog.

Miacis proved itself to be a highly-developed and extremely adaptable animal, ever ready to fill any vacated niche in the faunal kingdom, and eventually two distinct groups emerged, known as *Hoplophoneus* and *Dinictis*. *Hoplophoneus* gradually developed extra-long canine teeth and grew a special flange on the lower jaw, designed to cover the teeth when the mouth was closed. The sabre-toothed tiger, evocatively named *Smilodon*, was one species of *Hoplophoneus*. Totally unrelated to the great tigers of today, *Smilodon* was an immensely powerful cat about the same size as the present-day lion, with six-inch curved canines and a huge, hinged lower jaw which could open at right angles. This great beast was particularly skilled in seizing the heavy herbivores of its time as they slowly browsed and grazed in the fern forests and across the open plains. With its strong bones and massive build necessary to hunt down such large prey, *Smilodon* was doomed to extinction along with the great grazing animals, for its specialized skull structure could not be adapted fast enough for it to be able to pursue mammals which were evolving to become smaller and swifter and more specialized. Some interesting fossil remains dating back to the Pleistocene period of about one-and-a-half million years ago were discovered in a Brazilian cave and included in the find was a well preserved skull of *Smilodon*, proving that the creature prospered for several million years before its eventual demise.

The second group of *Miacis'* descendants flourished and sub-divided during the Miocene period of 20 million years ago. Known as *Dinictis*, this group gave rise to about 90 species of which some 35 still exist today in recognizable forms. All these species were directly related to the domestic cat, and one of the most successful of all the sub-divisions was a lynx-like animal with a neat head, forward-placed eyes and a streamlined muscular body. Its long powerful legs had small feet in which were sheathed strong retractile claws. This newly evolved trait enabled the cat to run swiftly on tiptoes, and to jump high, long and accurately while the claws were retracted. Then, having chased and marked their prey they could grasp, hold and tear with the

Twin kittens with 'wild-type' agouti coats

efficient weapons extended. The hooked claws also enabled the new hunters to climb high into the canopy of the dense forest trees whenever it was necessary to escape or hide from their own enemies.

Some species of *Dinictis* closely resembled the civet tribe of today but had rather shorter tails and longer, more pointed snouts. It is interesting to compare other members of the civet family such as the mongoose and genet with the fossil remains of *Dinictis* from the Miocene period.

Ten thousand years ago, in the Holocene period, men began to farm the land, cultivating plants for their own use and domesticating animals. Cave paintings of this period show all manner of animals in hunting and more domestic scenes, but in none of them are there any creatures remotely resembling our domestic cats. It must have been at this time, however, that the first timorous tabbies hungrily foraged for food scraps around human settlements, and found comfort in the warmth of the constant camp fires. What remains the greatest mystery of all, however, is the reason why these timid, furred creatures were not stoned or clubbed for the cooking pot. Instead they must have been actively encouraged, and the first links were forged in a chain of friendship that was to strengthen over the following centuries. No authenticated fossil records exist to prove the pathways taken by cats from the extinction of *Smilodon* until the well-documented emergence of the revered and deified cat of Ancient Egypt as recently as 5,000 years ago. The intervening years during which the cat assumed its true domestic role along with the dog, horse and other animals are as mysterious as the enigmatic creature itself, but by the time of the great Pharaohs, cat and man had established their firm, unique relationship.

THE CAT IN ANCIENT EGYPT

The cat was well established as a domestic animal in Ancient Egypt. In fact the Egyptians domesticated several types of cat, and some of our present day breeds and varieties may be descended from these deified creatures. Contemporary tomb paintings, frescoes, drawings and satirical cartoons depict Egyptian cats of elegant, slim conformation and often with large, pointed ears. The coat pattern, when indicated, is generally of spotted or mackerel tabby, with tawny or black markings on a lighter ground. Mummified remains of cats of this period have been identified as being of two main species which are still found today in their wild state, in areas of Africa. One is *Felis chaus*, the Jungle Cat, a fairly large, lynx-like cat, probably used for wildfowling and to kill rats in the granaries. The other is *Felis lybica maniculata*, the Gloved Cat, more easily tamed and most likely to have been kept as a housepet, and to guard the home against rodents and snakes.

One large wall painting, now in the British Museum, shows an Egyptian family enjoying a punting expedition into the marshes. The mother and daughter are depicted serenely gathering lotus blooms from the water, while the father propels the punt forward, putting up a crowd of water birds. The family cat, a large striped tabby, grasps small birds in each paw and another larger fowl is gripped firmly in his jaws.

A cult developed for the worship of cats in Egypt and lasted for more than 2,000 years. At first the Egyptians considered the cat sacred only to the Goddess Isis, then gradually the great Cat-Goddess Bastet emerged. She was sometimes referred to as Bast, or Pasht, and it is said by some that our word 'puss' derives from this revered name. As Bastet's animal incarnation was the cat, feline models were made in every conceivable form and every possible material. Tiny amulets were formed from precious stones and metals, larger figures were made of stone, glass, quartz, marble and all manner of metals. Cats were moulded in clay, carved in wood and fashioned in bronze.

Perhaps the earliest portrayal of Bastet as a cat-headed figure is on a papyrus of the 21st dynasty, now in the Museum at Cairo. At that time she took precedence over all other deities, and a great temple built to her glory stood firm for centuries, even after the cult diminished. It was sited at Bubastis, east of the Nile Delta, and was visited by Herodotus, the Greek historian, during his extensive tour of Egypt about 450 BC, and described in his later writings. Built by King Osorkon II as a festival hall, in the centre of the city, the temple was surrounded on all sides by wide canals, fed directly from the Nile. It was constructed of red granite in the form of a huge square with stone walls at its heart surrounding a grove of tall trees, sheltering the shrine which held the great statue of Bastet. On one of the walls of the shrine a carved relief showed King Osorkon II endowing the goddess, and has the inscription 'I give thee every land in obeisance, I give thee all power like Ra'.

Holidays, pageants and feasts were held to the glory of Bastet, and her sacred statue was sometimes brought out of the innermost shrine of the temple so that she could enjoy the festivities at which she was honoured. One of the principal and most popular of the festivals was the annual ceremony at Bubastis to which thousands of people from all over the country would make pilgrimage. Leaving their homes in April or May, they would board barges and set sail down the great river. Carried by the stream, the vessels made slow progress and the voyagers passed the time in merrymaking, playing flutes and cymbals, singing, drinking wine, dancing, and making frequent stops at villages en route, taking on supplies and more passengers.

Some 70,000 people would assemble in Bubastis, and after much wine had been consumed, many animals were sacrificed to the goddess Bastet, before feasting and street celebrations commenced. Then the great statue of the deity was carried out of her shrine and conveyed by barge to take her part in the festival. All the worshippers and local townspeople took enormous trouble in preparing the area and its decorations, and rehearsed their music and dancing to be of the very highest standard to please their goddess and invoke her blessings.

During Bastet's long reign, every cat was venerated. Housecats were loved and given delicacies from their owners' tables. They were petted and fussed, jewelled collars were put around the necks of rich men's cats, and some had their ears pierced to carry gold rings, studs or precious stones. Wild and untamed stray cats were also cared for, and food was regularly left in their haunts. Sick and injured cats were tended with the solicitude generally given to small children, and if a cat died the whole family entered a period of mourning. This entailed shaving off the eyebrows as a mark of respect, beating gongs

and wailing. The dead pet was mummified before being buried with full honours. Its body was expertly treated with precious oils and wrapped in layers of cloth by those skilled in embalming methods. Even quite poor people went to some expense to ensure that their cats were properly preserved, and rich owners made very elaborate funeral arrangements. In preserving a cat, the embalmers paid particular attention to the head: each ear was treated separately, and arranged to point upwards, while the four legs were bound tightly in with the body wrappings.

Mummy cases were made in all manner of materials. Small wooden coffins were common, and ranged from simple boxes to elaborately carved caskets. Simple wrappings of woven straw gave some mummified cats the appearance of wine bottles, with the clearly recognizable head at the top. Other cats' bodies were placed in beautiful bronze boxes, with cat figures or statues on the lids. Many such embalmed cats were taken or sent to special cat cemeteries for burial, including one at Bubastis, and at the turn of this century another site was discovered in central Egypt. This ancient burial ground, at Beni Hassan, contained many thousands of mummified cats. Not fully realizing the importance of the find, the excavators allowed tons of earth, rubble and embalmed cats to be carted away for use as fertiliser. Luckily, later finds were examined by experts who understood their importance, and it is from such preserved specimens that we have obtained much of our present knowledge of the cat's place in Ancient Egyptian culture.

As protectors of the granaries and the hearth as well as the incarnations of powerful gods revered by all men, it is little wonder that cats were jealously guarded and protected by their Egyptian masters. For centuries their sale or export was forbidden, but gradually, by devious routes, for gifts and as exchange, cats were taken by Egyptian monks and Phoenician traders eastwards into the Orient and westwards to Europe.

THE DISTRIBUTION OF DOMESTIC CATS

At the time of their worship in Egypt, cats of similar

type and temperament were being established in China and India as semi-wild pets, used in vermin control.

The Ancient Greeks are known to have kept cats to replace weasels in pest control, but they do not seem to have appreciated them as pets, although they certainly kept dogs and caged certain other small creatures such as cicadas. Only one piece of contemporary pottery shows a cat, being led by a slave.

Trade between Egypt and Rome resulted in the introduction of cats to Italy, where the grain stores had been previously protected from rodent attack by small carnivores, such as stoats, weasels and mongooses. A mosaic of about 100 BC was excavated at Pompeii. Perfectly preserved by the ash that fell in that great city's destruction in AD 79, it shows a beautifully executed cat with a large bird in its jaws, but there are no other signs to determine whether it was a pet animal, or a wild cat engaged in catching its natural prey.

The Romans were impressed by all things Egyptian and it became fashionable for wealthy households to keep pet cats. Wherever the Roman legions marched, they carried their precious cats. Occasionally, some were traded, and if the armies settled for several months, their cats possibly interbred with the local wild cats.

As the cat spread across the world its name accompanied it. It had been known as Mau, Mau-Mai, Maau or Maon by the Ancient Egyptians, from their word meaning 'to see' and probably referred to the animal's connection with the *utchat* or sacred eye of Ra. Our name 'cat' however, probably came from the Nubian word *Kadis*, which spread northwards from Africa in its related forms through the countries bounding the Mediterranean, the Baltic and the Atlantic. In most of these countries, cats are known by very similar names. In Holland and Denmark the word is *kat*; in Germany *katti* or *katze*; in Sweden *katt*; in Portugal and Spain *gato*; in Italy *gatto*; in Poland *kot*; in Russia *kots*; in Turkey *keti*; in Wales *cetti* and in France *chat*. In various other countries of the world, the animal is known variously as cath, catus, cattus, cait, katte, kottr, kazza, kate and kotu. It is certain that the cat had become widely established as a true domestic animal

at least 2,000 years ago for in the writings of Pliny the Elder, in the first century AD, his records relating to the cat, in all its forms, are accurate observations of the animal as we know it today, and never refer to any other wild or domestic animal or related species such as the mongoose, weasel or civet.

It is recorded that in AD 999 a cat gave birth to a litter of five kittens in the Imperial Palace of Kyoto, Japan, and the Emperor was so fascinated by this event which took place on the tenth day of the fifth moon, that he ordered special treatment for the feline family. Precautions were taken to ensure that the cats could not leave the palace, so that further generations of similar type could be bred.

At that time, cats were highly prized in Japan and many were kept in the silkworm farms to kill the mice that threatened to eat the developing cocoons. Following the Emperor's edict, however, it became fashionable to tether cats on silken leashes, controlling their breeding and wanderings. They were given the best of food and care and soon became sleek, fat and rather lazy. As the cats' natural hunting pursuits were curtailed, mice reached plague proportions and the silk industry approached the point of collapse. The granaries were also overrun with rats and mice and eventually in 1602, the government of the day decreed that all cats must be set free to hunt as they wished and fines were imposed on anyone found buying or selling cats. The versatile animals soon proved that the years of confinement had not impaired their hunting abilities – the rats and mice were soon brought under control once more and the silk farmers regained their markets.

Legislation pertaining to cats was also introduced to Britain during the tenth century. In the year 936, Hywel Dda or Howell the Good of Wales drew up many laws including those for the protection of cats.

The Law of Hywel Dda, King of South Wales AD 936
The worth of a cat and her tiethi is this – The worth of a Kitten from the night it is kittened until it shall open its eyes is one legal penny. And from that time it shall kill mice, two legal pence.

A typical feral tom cat with torn ears and battle scars

And after it shall kill mice, four legal pence, and so it shall always remain.

Her tiethi are to see, to hear and to kill mice, to have her claws entire, to rear and not to devour her kittens, and if she be so bought and be deficient in any one of these tiethi, let one-third of her worth be returned.

The worth of a cat that is killed or stolen is determined thus: let its head be put downwards upon a clean and even floor with its tail lifted upwards, and thus suspended whilst wheat is poured about it until the tip of its tail be covered – and that is to be its worth.

The worth of a common cat is four legal pence.

Whoever that sell a cat is to answer for her not going a caterwauling every moon, and that she shall devour not her kittens, and that she shall have ears, teeth and nails, and being a good mouser.

As various European countries developed their individual styles and cultures, the cat's role changed from that of a mere controller of vermin and it was often used in pagan rituals and magical rites. As cities grew and overcrowding became more common, the rat population also increased, devastating food stores and carrying disease. Wherever man went, the rat followed, its gradual spread tracing the pathways of invasion, exploration and trade. This pest was responsible for spreading the Black Death that ravaged Europe and Asia during the fourteenth century. In 1660, the black rat, living and breeding in the sewers of London, introduced bubonic plague which decimated the city's population. The cat was Man's only ally against the disease at this time, and its value soared. Despite this, the church outlawed the cat, because of its connection with witchcraft, and for the next few generations the unfortunate animal was encouraged as protector and friend on the one hand, and persecuted as an incarnation of the devil on the other.

As well as bringing back plague and pestilence from the East along with the rats in the holds of their ships, the Crusaders returning from the Holy Wars also brought many rare and unusual cats. These were chosen for their strange colours and a few of them had long fur. Some cats brought to Britain from Angora, now called Ankara, in Turkey, in the sixteenth century were reported as being 'Ash-coloured, Dun and Speckled cats, beautiful to behold.' They were described as having long, silky fur, small heads and long, pointed ears. Other long-coated cats were introduced from Persia, now called Iran, and these were of different bone structure, being generally of heavy build and with thicker fur, broader heads and shorter, bushier tails.

In the eighteenth century, the versatile and voracious brown rat spread across Europe to Britain, and even to the New World. Cats were the only efficient means of controlling it, and were employed as rat-catchers in stores, shops, warehouses, factories, farms and on board merchant ships. It was the introduction of cats on board ships that accelerated the spread of the species throughout the civilized world, for kittens were often born at sea and when independent, were likely to disembark in port, far from the land of their conception. Most seamen like cats, and will enjoy telling yarns of the exploits of their own ships' cats and their excursions abroad. The skill of cats as rat-catchers aboard ship was even recognized by the courts, and some maritime insurance companies would only indemnify for rat damage on a ship if there was evidence to state that cats were carried while the vessel was at sea.

The cat gained even more popularity in the middle of the nineteenth century when Louis Pasteur discovered the microbe. Dogs which had long been popular as pets were suddenly considered far from hygienic and were banished from many homes, while the dainty, decorous cat, always so punctilious about its toilet, was considered far safer to allow in the home, and to play with the children.

Despite its air of independence, the cat allowed itself to be accepted once more, like its Egyptian ancestors, into the hearts and homes of its admirers, first as a fireside pet, and in later years as a valuable acquisition, as specific breeds were developed and the Cat Fancy was born.

One of the most expert early naturalist writers on the subject of cats was Pocock (1863–1947). He regarded the study of the domestic cat as a scientific exercise and made some profound observations which stand up to this day. Pocock considered the patterning which we call tabby to be the wild-type coat of the earliest domestic cats, and described two tabby patterns, the first being the blotched design *catus* that we know as classic, or marbled, and the second being the striped or spotted pattern, *torquata*. He considered that *catus* had at some time abruptly arisen as a mutant form of *torquata*.

In the striped pattern (*torquata*) the markings are formed of narrow lines transversely or vertically arranged on the sides of the body, and may be solid or broken into shorter stripes, or elongated spots particularly towards the posterior of the animal. In the blotched variety (*catus*) the broader stripes loop and spiral behind the shoulders and three long stripes run dorsally to the root of the tail. In both types of tabby the tail is ringed and the face is boldly etched with distinctive lines including an 'M' mark on the forehead. Pocock explained that the two pattern types are individually variable and that one or the other could be detected in the skins of all European domestic cats, even though the stripes of the body might be entirely suppressed in the *torquata* phase, which he considered less stable than the *catus* phase.

Pocock also recorded some varieties of cats which had arisen by mutation and cross-breeding, including a tortoiseshell from Spain, blue-grey cats from Europe and Siberia, red cats from the Cape of Good Hope, Malayan cats with short twisted tails, a Chinese cat of black and yellow with folded ears, a black cat from South Africa, a greyish cat from Abyssinia, a piebald cat from Japan, a fawn and black marked cat from Siam and a jet black cat from Russia. From all these colours, together with the mutation which produced long fur, it is easy to understand how the many beautiful varieties of the present day have developed.

Feline Characteristics

The cat is a typical carnivore, being a flesh-eating animal which suckles its young, and has very specialized dentition adapted for stabbing, biting and tearing. It is a successful mammal, designed for survival in almost any environment: it can run swiftly, stalk efficiently, climb strongly, leap high and wide and hold firmly onto its prey.

Like all other living creatures, the cat's body is composed of cells so minute that they can only be seen with the assistance of a powerful microscope. While each cell is a complete unit of life in itself, specialized cells are grouped throughout the body, working together to perform different functions, and forming a variety of tissues and organs.

FRAMEWORK

Though it is so much smaller than Man, the cat has 230 bones compared to the 206 in the human body. The skeleton consists of various bony structures forming a semi-rigid framework designed to support the softer body tissues. The bones of the spine, chest, pelvis and limbs form a system of levers, operated by their attached muscles, while other bones such as the skull, rib-cage and pelvis, protect vital organs.

Bones are of four distinct types. *Long bones* are roughly cylindrical, with hollow shafts containing bone marrow, the medium in which precious red blood cells are manufactured. Such bones – the femur, the humerus, the tibia, the fibia etc. – form

Silver Tabby Persian – Champion Karnak Mailoc

the limbs. *Short bones* have a spongy core surrounded by very compact tissue and are found in the feet and kneecaps, while *irregular bones* are similar in structure but of variable shapes, and form the spine. *Flat bones* are made from two layers of compact bone with a spongy layer sandwiched between and are found in the skull, the pelvis and the shoulderblades.

The skull is made up of a number of pieces of flat bone, which are sectioned, rather like a puzzle. In very young kittens, the pieces are not completely

15

joined, so great care should be taken to prevent head injuries. The flat bone sections are pierced by tiny holes threaded with nerves and blood vessels.

The long string of irregular bones forming the spine is attached to the skull at one end, and terminates at the other in the tail tip. The hollow sections of the cervical, thoracic and lumbar vertebrae contain the vital spinal cord, and bony projections on each of the vertebrae form attachment points for the strong back muscles. The amazing flexibility and agility of the cat is produced by the design of the connections between the spine and joints at the shoulders and pelvis.

Flattened, elongated bones form the 13 pairs of ribs and contain bone marrow; the attached muscles control the cat's breathing by causing the lungs to empty and fill involuntarily. Flat bones in the pelvis are fused together in pairs, forming a girdle, and the scapula or shoulder blade consists of flat bone of triangular shape. Free movement is possible in the cat's shoulders, as, unlike Man, it does not have a fixed collarbone. The *humerus* is the first long bone of the arm, and fits into the shoulderblade. At its lower end the humerus joins with the thick *radius* and the thinner *ulna* bones which are parallel and form the forearm, the part of the foreleg which stands free from the cat's body. The forepaws have sets of three small bones, forming individual digits, each of which has articulated end bones, allowing the hooked claws to be extended or contracted.

The hindlimb has a very long *femur* or thigh bone, attached to the pelvis by means of a ball and socket joint, and jointed to the *tibia* at its lower end, forming a kneecap, or *patella*. From the patella, the hindlimb is free from the body and the long strong tibia is reinforced by the slender *fibula* from kneecap to ankle or hock. Though similar to the bones of the forepaws, those of the hind feet are longer, and the first toe is absent. The cat's typical graceful, gliding and flowing movements are produced because it walks like a ballerina, poised always on the tips of its toes.

Under normal conditions, the cat's skeleton affords adequate protection and support, but accidents can cause fractures and dislocations. Even severe fractures may be mended by skilled veterinary treatment, although the cat is a notoriously bad patient, and resists the rest necessary for convalescence. When one or more of the bones forming a joint become displaced, a dislocation occurs and urgent treatment is necessary to replace the joint before severe swelling sets in. Some cats suffer regular attacks of dislocation of the kneecaps, and as this is thought to be the result of a hereditary condition, they should never be used for breeding further generations.

There are few examples of skeletal abnormality in the cat, for the species has not been subjected to as much selective breeding as its canine cousins. Defects sometimes encountered include kinked, bent or shortened tails, cleft palate, flattened chest, polydactylism and split-foot.

A twisted or bent tail may be caused by injury at birth or in later life, or it may, it is thought, be caused by an event during the development of the kitten, such as trauma experienced by the queen, or antibiotic treatment. Shortened tails are generally due to the presence of a simple recessive gene while kinked tails result from a fairly complicated genetic pattern.

Cleft palate is often associated with harelip, and the defects arise because of the failure of the two halves of the hard palate and upper lip to fuse at the normal time during the gastrolation process in the embryo kitten. The conditions have occurred as the result of the administration of certain drugs to pregnant queens, and have also been found to arise in certain breeding lines, indicating a hereditary cause.

Flattened ribcages and other skeletal defects in kittens sometimes occur, particularly in those born to queens with a deficiency of Vitamin A in their diet. Polydactylism is the presence of extra toes on the paws and is fairly common in some cat populations, being caused by a hereditary factor. The number of extra toes varies considerably, and some cats may even appear to have double feet on each leg. Conversely, in the split-foot condition, toes of the forepaws are generally missing, or may be fused together, often with a long cleft in the centre of the paw. Some of the toes may have double claws, and this condition can severely affect the mobility of the afflicted cat. Split-foot is caused by the action of a dominant gene, so affected cats should be neutered.

MUSCLES

A complex system of muscles gives the shape of the feline frame, and controls its movement by alternate contraction and relaxation. Those muscles that straighten and extend a joint are called *extensors*, while those used for bending or flexing a joint are known as *flexors*. *Abductors* are muscles which move the limbs from the body and *adductors* draw the limbs to the body.

Muscles are made from specialized fibres which contract in response to signals from the nerves, and the cat has three muscle types. *Striated* or *striped muscle* is attached to the limbs and other parts of the anatomy under voluntary control while *smooth muscle* carries out involuntary movements such as those of the intestinal wall and the blood vessels. *Cardiac muscle* is adapted for carrying out the important functions of the heart and possesses special powers for maintaining rhythmic contractions.

CIRCULATION

The blood of the cat passes around its body in a typically mammalian double cycle. The heart first pumps the blood through the lungs where it is oxygenated, then pushes it around the rest of the body. Strong arteries carry blood from the heart and that efficient organ's pressure pushes it through fine capillaries in the tissues. Here gases, nutrients and hormones are exchanged. Depleted blood is collected by veins and transported back to the heart for recirculation. In the cat, the circulatory system is adapted to allow rapid changes from rest to extreme exertion and if the blood picture is marred in any way, the animal soon becomes noticeably ill. Heavy infestations with intestinal parasites quickly destroy large numbers of the cat's red blood cells and produce anaemia, and some drugs may prove toxic to the cat, destroying bone marrow and preventing production of sufficient new red blood cells. Aspirin should never be given to cats as it has the effect of destroying circulating cells in the bloodstream itself.

SKIN AND HAIR

The skin of the cat is made from two layers of tissue each consisting of collections of specialized cells. The outside layer of skin is called the *epidermis*, and is being constantly replaced as the surface area dies and sloughs away. The inner layer is known as the *dermis*. In a healthy cat, the skin is pliable and elastic, while in the sick animal it may feel hard and stiff. Its colour, too, may indicate the animal's physical state. A pallid skin could point to infestation with parasites, shock, or lack of some vital nutritional component in the diet. A reddish skin indicates inflammation, while a bluish tone could mean respiratory disease. A yellow tinge to the skin generally means that the cat has jaundice, which is one symptom of several serious feline diseases.

Sweat glands in the skin excrete impurities from the body, but seem to play very little part in controlling the cat's temperature. Sebaceous glands open into the hair follicles and secrete sebum, an oily substance which coats each hair as it grows from the skin surface. The pads and nose leather of the cat are free from hair, and while the pads are very sensitive to pressure, the nose is highly sensitive to touch. The skin of the pads is tough and pliable and secretes sweat when the cat is hot or alarmed. The sharp, hooked claws grow continuously from their sheaths, and are formed of sheets of keratin, just like human fingernails.

Hair grows as an insulative layer, forming a dense pelt over the body, keeping the cat warm in cool weather and cool in warm weather. Each individual hair is a long thin, cylindrical structure, pointed at one end and with a tiny bulb at the other, which is embedded in the dermis. Each hair is formed separately in a follicle within the skin and as it grows, pigment cells inject it with its genetically determined colour. Special muscles are attached to some follicles, enabling the cat to erect its coat when it is angry, startled, or has difficulty in adjusting its body temperature. In certain areas, the hairs are drastically modified to produce features such as the whiskers. The *vibrissae*, as they are called, are very sensitive and the slightest touch to the tip sends a message through a nerve in the upper lip to a larger nerve which transmits directly to the brain. Whiskers are used for sensing environmental changes, the width of openings, and to signal emotion.

Moulting is the term applied to the shedding of dead hairs, and usually takes place at certain seasons of the year, producing typical summer and winter coats. Cats kept in controlled temperatures often shed throughout the year, their coats remaining fairly constant regardless of the outside weather. Noticeable loss of hair may occur during the course of some illnesses such as eczema, mange and ringworm, after poisoning, and following some hormonal disturbance, and its cause should be determined without delay.

EYES

Humans are said to have better daylight vision than cats, but after dusk the cat scores, for although it cannot see in total darkness, its pupils expand to give excellent vision in the dimmest of conditions. In the cat's eye, light passes through the curved *cornea* and *lens* to the *retina*, to be reflected by a special layer of iridescent cells known as the *tapetum lucidium*, which is thought to have a photomultiplying effect on the accepted light. In strong daylight the pupil contracts to a narrow slit, protecting the delicate mechanism at the back of the eye, while still allowing adequate vision. Both eyes face forward, allowing the fields of vision to overlap, producing stereoscopic sight and enabling the cat to be extremely accurate in its judgement of distance, a great asset in hunting. With large orbs, deeply set in the skull, the cat's eyes cannot move very freely within their sockets, and the animal often needs to turn its head or body in order to focus sharply.

As well as the normal upper and lower eyelids, the cat has a third lid called the *nictitating membrane* or haw, a pale skin-like structure normally kept out of sight inside the inner corner of each eye. Its function is to remove dirt and debris from the eyeball, and it automatically sweeps across the orb when the outer eyelids close.

Sick cats often have the haws showing across the inner corners of the eyes, due to the shrinkage of the little pad of fat which normally holds the eyeball forward in the socket. In fact the appearance of raised haws in an otherwise healthy cat points to

something being amiss, and although the symptom could be mild, it is always worth taking veterinary advice.

EARS

The ear of the cat may be considered in three sections. Outside is the erect earflap, or *pinna*, which normally points forward, but is flexible enough to move sideways and back in order to pick up and determine the direction of very slight sounds. Sound waves are caught by the pinna and funnelled down to the *eardrum* which stretches across the ear canal. The middle ear has three small bones which transmit the sound waves to the inner ear, where they are analysed and converted into impulses and passed along the acoustic nerve to the brain. In the brain, the auditory cortex decodes the signals, recognizing familiar ones by comparing them with sounds stored in the memory bank. This fairly sophisticated system enables the cat to learn many sounds, and to recognize voices and its own name.

Hearing is very well developed in the cat, and it can register frequencies two octaves higher than those detected by the human ear, although it is less sensitive to sound of lower frequencies. High-pitched squeaks, inaudible to humans, could indicate a welcome meal for the feline hunter, and the efficiency of its auditory system combined with its capacity to see well at dusk have contributed greatly to its success as a wild species.

Hearing acuity diminishes with increasing age, and many geriatric cats are deaf. Deafness may also result from ear disease, or as a side effect of some drugs. It can also be caused by hereditary factors, and is linked to one form of white coat colour in cats, especially when such cats have blue eyes. The genetic factor which causes the whiteness of coat and the blue of the eyes is linked with changes in the *cochlea*, the inner ear, which are severe enough to inhibit passage of sound waves.

Ear flaps are vulnerable to injury, and are often torn in fighting. They are inclined to bleed profusely, but heal quickly, and unless the underlying cartilage is damaged, regain their original appearance. If blood blisters develop, or the cartilage is torn, the ear may heal in a contracted or folded manner.

Tiny mites may invade the inner passages of the ear, breeding and feeding in its moist warm lining. Such parasites can lead to serious infections, or, if neglected, to deafness, as well as causing intense irritation to the unfortunate afflicted cat.

NOSE

The cat's sense of smell is highly developed and is stimulated by minute particles of odorous substances in the air, either drawn in with the natural breathing process or deliberately sniffed. In the nasal cavities sensitive nerve endings form fine olfactory hairs, and these are linked to nerve cells connected to the brain. The olfactory centre in the brain of the cat is much larger than expected in so small an animal, indicating the importance of the sense of smell to the species. As well as assisting the animal in its hunting exploits, smell is an essential part of its sexual life. A small pouch lined with receptor cells is situated at the back of the roof of the cat's mouth. Known as the *Jacobson's organ*, this is used for the identification of subtle scents, and a cat that is trying to identify a special smell will often hold its mouth half open in an odd grimace, called a *flehmen reaction*.

Cats will not eat carrion and find even slightly tainted food distasteful. They often seem to appreciate perfumes, especially those manufactured on a musk base, and will rub their heads ecstatically on a particularly attractively scented area.

Certain plants emit odours which attract cats, and catnip or catmint *Nepeta cataria* is found by some felines to be totally irresistible. The ability to detect and respond to catnip seems to be inherited by about 50 percent of the cat population, and those that appreciate the scent enter a state of extreme pleasure which lasts for about ten minutes. The characteristic behaviour pattern is for the cat to approach and sniff the catnip, then to chew the leaves, rub the head and body over the plant and roll ecstatically in it. Tom cats tend to become more excited by the plant than females, and the response varies in intensity between individual cats.

Regardless of their sex, cats often respond to the scent by posturing and treading, rather like the female cat in oestrus, but this response is probably unrelated to sexual behaviour because neutered cats are affected just as much as entire animals. Catnip grows as a wild plant through the temperate regions of Europe and North America and the substance that affects cats has been extracted and analysed. *Nepetalactone* is commercially available as a powder or spray to be used on cat products and toys. It seems likely that a psychedelic state is induced by *nepetalactone* molecules reaching the brain centres. Some affected cats stare raptly into space, while others chase imaginary prey. The effect of catnip is quite harmless, of very short duration and non-addictive. Some other plants (such as the Valerian, *Valeriana officinalis*) attract cats too, but rarely have the intense effect of catnip.

MOUTH

Mucous membranes line the inside of the cat's mouth and are kept moist by the action of salivary glands which wash away bacteria and dead cells, as well as softening food to allow easier swallowing. The tongue of the cat is very specialized, and rough due to the *papillae* on its upper surface. These are of several kinds, and those along the centre of the tongue form backward-facing barbs able to hold food or prey, and to lick meat from bones. There are mushroom-shaped *papillae*, complete with taste receptors, along the front and sides of the tongue, while cup-shaped *papillae* are found right at the back. The tongue is used for grooming the cat's own fur and for social grooming, as well as for rasping and softening food, and its edges can curl, allowing efficient lapping of liquids.

Cats are unusual pets in that they show no liking for sweet things, but do have high sensitivity to the taste of water, and easily differentiate between salt or non-salt liquids. Cats soon learn to tell the difference between various meat smells, and can form definite preferences and dislikes for certain foods.

Truly carnivorous, the cat has highly specialized dentition with 12 incisors, four canines, ten premolars and four molars present in the full adult mouth. Kittens are born with teeth just visible in the gums; these slowly erupt, and can be felt like tiny needles by six weeks of age. The baby teeth are shed as the permanent teeth come through and the complete set should be established by the time the kitten is seven months old.

DIGESTION

The teeth of the cat are not designed for chewing, but for biting and tearing. In its wild state, the cat first catches its prey, tears a lump from the carcase, then swallows the piece whole. Salivary juices in the mouth have little time to start the processes of digestion and any parts such as feather or hair, which are not quickly broken down in the stomach, may be regurgitated. The stomach juices of the cat are very strong and capable of reducing even hard bone into a cartilagic state. In the stomach, proteins start the long process of conversion necessary to produce the exact amino acids required to regenerate parts of the cat's body. Partly digested food is passed through the pylorus into the *small intestine* where further digestive processes take place, helped by secretions from the *pancreas* and *liver*. Fats are broken down and extracted, sugars are altered structurally so that they may be stored, and minerals are absorbed. Fluids pass from the small intestine into the *large intestine* to be attacked by specialized bacteria, water is passed to the *bladder* and wastes pass into the *colon* and are then evacuated.

In humans, salivary glands produce a powerful enzyme *ptyalin*, which begins the process of breaking down starches, so that they can be converted into blood sugars. This enzyme is rarely present in feline saliva, resulting in starches passing virtually unaltered into the small intestine. When this happens, diarrhoea may result, and many cat owners increase the cat's intake of carbohydrates in an attempt to add bulk, but in effect this only aggravates the situation. There is little point in feeding starches to a true carnivore.

CONFORMATION AND COAT LENGTH

Modern pedigree cats come in two main conformation types, one which is large boned, short-coupled and stocky, with a rounded head, small ears and a short nose, and the other which is fine-boned, long and rangy, with a pointed head, large ears and a long nose. Most of the long-coated breeds, and the British, American and European Shorthairs, have conformation of the first type, while some of the semi-longhaired varieties and the Foreign or Oriental Shorthairs including the Siamese, all have conformation of the second type. Some breeds classed as Foreign Shorthairs fall between the two extremes and must conform to strict standards of points laid down by their breed societies and ratified by the governing bodies. Many so-called breeds are in fact merely colour varieties within the same breed, and while the cats must be of the same general body type, the individual colouring and eye colour must conform to the desired standards. In this section I shall explore the main categories of Fancy felines, then follow the genetic pathways which have produced the many varied cats which grace the show benches of the cat world.

Persian and Longhaired Cats

Today's pedigree longhaired cats are the descendants of two distinct types which arrived in Europe during the sixteenth century.

The first to arrive was the Angora, brought from Ankara in Turkey by the writer Nicholas-Claude Fabri de Peiresc. The cats were said to be 'ash-coloured, dun and speckled . . . beautiful to behold'. They had long silky fur, long noses and fairly large ears. Later, stockier longhaired cats were brought from Iran, and the two types were bred quite indiscriminately together. By the end of the eighteenth century, the heavier Persian type had superseded the lighter framed Angora conformation, and selective breeding began for larger bone and heavier coats.

In 1868, Charles Ross produced a book about cats and described the differences then seen between the Angora and the Persian. Of the former he wrote:

. . . a very beautiful variety, with silvery hair of fine

silken texture, generally longish on the neck, but also on the tail. Some are yellowish, and others olive, approaching the colour of the lion; but they are all delicate creatures and of gentle disposition.

Of Persian cats he wrote:

a variety with hair very long and very silky . . . it is differently coloured, being of a fine uniform grey on the upper part, with the texture of the fur as soft as silk and the lustre glossy; the colour fades off on the lower parts of the sides and passes into white or nearly so, on the belly. This is probably one of the most beautiful varieties and is said to be exceedingly gentle. . . .

In 1871, Harrison Weir organized Britain's first official cat show at London's great Crystal Palace exhibition centre, and this was such a resounding success that it formed the foundation stone of today's flourishing Cat Fancy. Further cat shows were held in Britain, and then in other parts of the world, and official standards or points of perfection were drawn up.

The National Cat Club was formed in 1887, and became the first registering body in the world for the keeping of official cat records. Persian cats proved the most popular of all breeds, perhaps because they were owned by many members of the aristocracy. Queen Victoria owned two Blue Persians and set the royal seal of approval on the Cat Fancy by attending shows and offering special awards.

By the time the Governing Council of the Cat Fancy was established in 1910, there was a wide range of accepted longhaired cats, some of which had their own specialist clubs, such as the Blue Persian Cat Society, the Black and White Cat Club, the Chinchilla Cat Club, the Silver and Smoke Cat Society and the Orange, Cream, Fawn and Tortoiseshell Society. All these drew up their own breed standards and nominated judges.

Today, Persian and longhaired cats are grouped into self-coloured, non-self-coloured and semi-longhaired varieties and are bred specially to conform to exacting, approved standards.

Shaded Golden Persian – Gray-Ivy Golden Nugget

In the self-coloured gtoup there are old-established varieties known as Black, Blue, Red and Cream, plus the White, which may be blue-eyed, orange-eyed or odd-eyed, where one eye is blue and the other orange or gold. More recent self-coloured longhaired varieties are the Chocolate and Lilac.

The non-self group is larger, and includes the sex-linked and usually all-female varieties such as the Tortoiseshell, the Tortoiseshell-and-White and the Blue-Cream. There are silver varieties such as the Chinchilla and the Smoke as well as the Shaded Silver, plus the Cameo range of Shell, Shaded and Red Smoke.

Tabby Longhairs may be had in a silver variety too, as well as the Brown Tabby and Red Tabby, but there are no spotted Persians.

There are strikingly marked Bi-colour Long-hairs, which may be of any recognized cat colour, with white, and a whole range of Himalayan-patterned colourpoints.

The Birman is another Himalayan-patterned longhaired cat, but does not have the same con-formation as the Persian and is therefore classed as a semi-longhaired variety. It is an ancient breed and distinguished by its four sparkling white feet.

Another 'pointed' cat with long fur is the Balinese, a long-coated Siamese cat, and there is a long-coated form of Abyssinian too, known as the Somali.

Perhaps the rarest of the semi-longhaired varieties throughout the cat world is the Turkish, originally called the Van Cat and said to enjoy swimming.

Longhaired cats are all very beautiful, but must receive regular attention to keep their flowing coats in the correct condition. Most varieties need a thorough combing every single day, or the under-coat forms into tight knots which become im-possible to untangle. They are gentle and un-demanding cats however, with quiet voices and affectionate ways. Each breed has slightly different characteristics, and some coats seem less inclined to knot than others.

British Shorthairs

At the first British cat shows, shorthaired cats were greatly favoured and outnumbered the rarer long-haired varieties. But this situation was destined to be short-lived.

The National Cat Club staged its show at the Crystal Palace in 1896, and for the first time, Persians were in the majority, starting a trend that has continued to this day.

In an attempt to increase the interest in short-haired cats, a concerned group of fanciers formed the Short-Haired Cat Club in 1901. Mrs Middleton of Cheyne Court, Chelsea, London, was the Hon-orary Secretary, and the annual subscription was five shillings.

The club's main aim was to encourage careful breeding of pedigree British Shorthairs and it drew up careful standards of perfection in conformation and colour, and allocated points for each character, totalling 100 per cat.

At the same time another group formed the British Cat Club, but this lasted only a few years, while the Short-Haired Cat Club, with few changes over the decades, exists today as the Short-Haired Cat Society.

The British cat of the nineteenth century was the short-coated descendant of the various cats brought to the British Isles by the Roman armies, Phoenician traders and seamen of other lands. Interbreeding occurred naturally and eventually a fairly consistent stocky and tough type emerged. Built for survival in an often inhospitable and inclement land, the British cat evolved with a rounded head, large forward-looking eyes, small alert ears, strong jaw and a short, dense and virtually weatherproof coat.

Cats of all colours were seen, and by the turn of the century the Classic Tabby and the Spotted were favoured, along with some of the selfs, including the Black and the White.

In her *Cats for Pleasure and Profit* (revised) of 1905, Frances Simpson discussed the shorthaired breeds, pointing out that the Black should have no white markings and its eyes must be orange, while the White must be also unmarked by any other colour and its eyes must be blue. She stated that 'the commonest of all cats are shorthaired Tabbies and Whites, and Blacks and Whites'.

The British Shorthair of today conforms to an exacting G.C.C.F. standard of points. It is a com-

A winsome Silver Tabby British Shorthair at the
Taishun Cattery

pact, well-balanced and powerful cat with a deep
body, a broad chest and a short neck.

Its short strong legs have rounded paws, and the
thick-based tail has a rounded tip. The large round
head has good width between small rounded-tipped
ears and the chin is firm and well developed. The
eyes of the British Shorthair are large and well
opened and are set wide apart. The nose is short,
straight and broad.

For show purposes, the British Shorthair is
generally split into two groups, the self-coloured
and the non-self-coloured. The first group covers
the three white varieties: the Orange-eyed White,
the Blue-eyed White and the Odd-eyed White, as
well as the Black, the Blue and the Cream which
have orange or copper eye-colour. In addition,
breeders are currently working on the development

of Self-red, Chocolate and Lilac cats of true British
type.

The non-self group is much larger and covers
Shorthairs with more than one colour in their coats.
There are three classic British Tabbies – the Silver
Tabby, the Brown Tabby and the Red Tabby – and
all must show the desired marbled pattern with its
clearly defined 'butterfly' and 'oyster' markings, an
'M' on the forehead and 'Lord Mayor's chains'
around the neck.

Mackerel Tabbies are also recognized, although
rare, and the Spotted has become popular in recent
years, especially in the silver variety.

There are three all-female varieties in the non-self
group, the Tortoiseshell-and-White, the Tortoiseshell, and the Blue-Cream.

Mates for the former are generally found among

Blue Mackerel Tabby American Shorthair – Satinsong
El Cajon

the various Bi-colour Shorthairs, which help in selection for the correct amount of white spotting, while the Tortoiseshell is generally crossed with a Red or Black Shorthair male, and the Blue-Cream with a Blue or a Cream.

Bi-colour British Shorthairs may be of any accepted colour and white, and should be symmetrical in design if possible, and with a white blaze down the face.

Smoke Shorthairs are attractive, with pale silver undercoats and yellow or orange eyes.

Recognized in both Black and Blue varieties, the Smoke is closely related to the most recent addition to the shorthair scene, the British Tipped, which resembles a short-coated Chinchilla – a white cat with a coloured tip to each hair.

American Shorthairs

The American Shorthair is believed by some naturalists to be descended from the original domestic cats landed on American shores. Adapted to domes-

ticity without losing its natural intelligence, it makes an ideal working cat. Lithe and powerful, with a body free from exaggerated features, and with perfectly controlled reflexes, it is a natural hunter. Its typically short, dense coat forms a perfect protection from injury and cold. The head is large, with a square muzzle, full cheeks and a firm chin. The medium-sized ears are set wide apart and have rounded tips and the round, wide-set eyes are clear, bright and alert. The well knit medium-sized body is muscular and powerful, carried on strong, medium-length legs, ending in rounded paws. Show faults of the American Shorthair include over-long or fluffy fur, a deep nose-break, a kinked or otherwise abnormal tail or the incorrect number of toes. Such cats are also penalized if their body type is either excessively cobby or rangy.

Foreign and Oriental Shorthairs

The group of cats classed as Foreign or Oriental Shorthairs differ greatly from the British Short-

Zoobs, a handsome Blue-point Siamese

hairs. They are essentially long, lithe cats of medium size, fine boned, elegant and svelte. The first cats of Foreign conformation to be accepted in Britain were the Siamese, brought from Thailand towards the end of the last century. Originally seal-pointed, they are now bred in every possible colour, including a range of tabby- and tortoiseshell-pointed varieties. From Siamese have been developed elegant self-coloured cats in a corresponding full range of colours and patterns, some of which are recognized for show competition, and some of which are still classed as Experimental varieties.

This group of cats also includes breeds of less extreme but uniquely identifiable type, each of which can claim its own long recorded history. Some are bred in a range of colours, including the Abyssinian and the Burmese, while the Russian and the Korat are always Blue.

The coat of the Foreign and Oriental Shorthair is of a different texture to that of the British Shorthair, being very fine, short and close-lying, except in the case of the Russian, where it is required to be of a plush texture. Most cats in this group have Oriental eye shape.

3

Colours, Patterns and Breeds 1

AGOUTI AND TABBY CATS

The wild-type coat of the domestic cat is known as agouti, and is a universal camouflaging coat pattern found in many species, helping them to merge with the background colouring of their natural environment.

The tabby patterns that have been selectively bred into our pedigree breeds are quite complex and basically consist of an overlay pattern of a pre-scribed type, supported by a true agouti base.

Various breeds are recognized in a range of tabby patterns: Ticked, Spotted, Mackerel and Classic. The clarity and correct distribution of the overlay pattern is determined by the cumulative effect of polygenes, or groups of minor genes, whereas the factor which determines whether or not a cat is tabby in the first place depends on the action of monogenes, major genes with clearly identifiable effects. Agouti, the gene which allows a cat to show its tabby character, is dominant to non-agouti, and that means that tabby kittens can result from matings where only one parent is tabby, and can never result from matings where neither cat is tabby – in fact, tabby cannot be 'carried'.

Genetically, the Ticked Tabby is dominant to Spotted and Classic, and Spotted is dominant to Classic. Indiscriminate breeding between tabby cats of diverse patterns can lead to problems, for as the polygenes go into action poorly patterned cats may result.

Tabby cats may be bred in several colours, some officially recognized and some not. In Shorthairs, the patterns show up more clearly than in Long-hairs, and some non-agouti cats may appear to be tabby – though they are not genetically truly so – when the natural coat colour allows ghost markings to shine through.

In America there is also a 'Patched' Tabby pattern, known as Torbie, and a wider colour range, including Blue, Cream and Cameo. Only Classic Tabby Longhairs are recognized in the G.C.C.F., while British and American Shorthairs accept Classic, Spotted and Mackerel. In the Orientals, only Spotted is accepted, although Classics are shown in assessment classes and many breeders have produced poorly marked Mackerels, and the occasional Ticked. Tabby has been added to other breeds too, notably the Siamese, with a range of tabby-pointed varieties, and is accepted in Colour-points. Breeders strive constantly to improve the patterning of their tabby varieties, but sometimes a walk round the non-pedigree section of a cat show will reveal the finest and most clearly defined coats of all, produced not by clever manipulation of genes, but by sheer chance and natural selection.

Ticked Cats

ABYSSINIAN
The most famous of all the Ticked Tabby breeds is the Abyssinian, a cat which has been selectively bred for many years, reducing the incidence of natural tabby barring which would normally be found on the cheeks, around the neck, on the legs and tail and under the belly. Today's show standard Abyssinian has been relieved of all such unwanted tabby bars

and its coat should be unadulterated agouti, each hair having two or three bands of darker colour, producing an overall glowing, ticked effect rather like that of a Belgian Hare.

The Abyssinian is an attractive and unusual shorthaired cat with conformation unlike that of any other Foreign variety. It is one of the oldest of the domesticated breeds, having been first officially recognized as such in 1882. Lovers of this breed remark upon its similarity to the bronze statuettes of the cats of Ancient Egypt, and many insist that the Abyssinian descends in an unbroken line from the sacred cats of the great Pharoahs.

An Abyssinian-type cat, Zula, was brought out of Abyssinia (now called Ethiopia) at the end of the war there in 1868. She travelled with a returning expeditionary force to Britain and was kept by a Mrs Barrett-Lennard, but it is not confirmed in written evidence whether or not this cat was the forebear of the Abyssinians of today.

By the turn of this century, the Abyssinian cat had gained in popularity and was seen at the early cat shows. John Jennings, in *Domestic or Fancy Cats*, 1893, says: '. . . whether imported or a manufactured cross hardly matters, as it now breeds fairly true to point . . . no variety has yet rejoiced in such varied names, several countries claiming it as their own'.

The breed had many names, being known as the Russian, the Spanish, the Hare Cat, the Bunny Cat, the Ticked and the Rabbit Cat, with some folk even insisting that it came from a cross between a rabbit and an ordinary domestic cat and thus gained its distinctive ticked coat.

Whatever its beginnings, the Abyssinian is a firmly established, popular breed. It is lithe and sinuous, but not so svelte as the Siamese. Its heart-shaped head has large, wide-set ears, often with lynx-like tufts at the tips. The large, bright and expressive eyes may be yellow, green or hazel in colour. In the Usual or Ruddy Abyssinian, the short, fine coat is a rich tawny brown, each hair having very dark bands of colour.

The Sorrel Abyssinian is similarly ticked, but its coat, produced by the elusive 'light brown' gene, is a rich, glowing, red colour.

It is important that the show Abyssinian is free from tabby bars or markings on its head, tail and chest, and although all tabby or agouti cats have lighter underparts, chins and throats, the Abyssinian is penalized for having any white markings except for a small area under the chin.

Some breeders are currently working on the perfection of other colours within the breed, and attractive Blue, Chocolate, Lilac and Silver Abyssinians have been displayed at some cat shows.

SOMALI

It is now accepted that some of the original Abyssinian cats carried the recessive gene for producing long hair, but as two such cats had to mate together before long-coated kittens could be born, this phenomenon happened only rarely, and was then treated as something of an inconvenience. Any long-coated kittens present in an otherwise normal Abyssinian litter were generally regarded as 'sports' and were given away to be neutered pets. Long-coated Abyssinians arrived in litters born in Europe, Australia, New Zealand, the United States and Canada, as well as in England, but it was the Canadians who had the foresight to develop the unusual cats into a separate breed.

Gradually, more and more of the long-coated kittens were born in North America, and the name Somali was chosen for the breed as a tribute to the alleged African birthplace of the Abyssinian.

The Somali Cat Club was formed in the United States in 1972. Mated together, Somali cats produce only Somali offspring, but mated to normal Abyssinian, all the kittens are short-coated and are classed as Somali variants. Each variant carries the gene for the long coat and may in turn, be mated with Somali.

A Somali Cat Club has been established in Britain for the protection and development of the breed, and a preliminary standard of points has been drawn up, differing very slightly from that approved by C.F.A. in America. The Somali should give an overall impression of being a well-proportioned cat of medium size and Foreign type, firm and muscular, lithe, lively and alert. It should be easy to handle, with an even temperament, and feel weighty for its size.

The head should be a moderate wedge shape, the

Usual or Ruddy Abyssinian out hunting – Taishun Abigail

The longhaired Abyssinian is called the Somali, represented here by an appealing kitten of the Rainkey line

brow, cheeks and profile all following gently curved contours. The muzzle must not be pointed, and shallow indentations define the nose and cheeks. The ears are quite large, well cupped, set wide apart and pricked. Typical tufts at the tips are a feature of the breed. The eyes, which may be either amber, hazel or gold and almond-shaped, are large, brilliant and expressive.

The legs should be in proportion to the body, and the oval feet have tufts between the toes. The tail is full and brush-like, thick at the base and tapering towards the end, again in proportion to the body.

The special coat of the Somali is fine, dense and double-textured. It is basically of medium length, long at the ruff and 'trousers' and shorter over the shoulders. Two colour varieties are accepted, the Usual, called the Ruddy in C.F.A. and the Sorrel which C.F.A. knows as Red.

The Usual colouring is a rich golden brown over the body, each hair with at least three bands of black ticking giving a glowing quality to the coat. There is darker shading along the spine and tail, and black tips to the ears and tail, plus black pigment outlining the eyes, and dark lines extending vertically from the eyes towards the ears. The underbody is a rich apricot and free from ticking, bars or spots. The feet have black or brown paw pads and markings between the toes, extending up towards the hocks of the hindlegs, and the tufts between the toes are also black or brown.

The Sorrel Somali is a warm glowing red, ticked with chocolate-brown, and where the Usual is marked with black or dark brown, the Sorrel is marked with chocolate-brown. The nose leather differs too, being tile red in the Usual and pink in the Sorrel, matching the pads of the paws. Like the Abyssinian, Somali kittens are born with heavy markings and these may take several months to clear, leaving the correctly ticked coat.

Spotted and Mackerel Cats

Geneticists still disagree on whether mackerel and spotted coat patterns are due to different genes, or are caused by the action of the same monogenes, affected by polygenes. As breeders of spotted cats will know, it is often the mackerel-striped youngsters that eventually have the clearest spotting as adults, though this is not always true. The best spotted cats should have the spots arranged on their coats in random fashion, so that even if they tended to merge together, it would be impossible for the solid markings to form straight lines. It is rare today to see a true Mackerel Tabby, which should have sharply defined vertical stripes running down the body from the spine-line. Like the wild type, the Mackerel Tabby has distinctive stripes on its cheeks, the 'M' and scarab mark on the top of its head and bars on its tail and legs, necklaces and marks on the belly.

As they are clearly shown on Ancient Egyptian papyri, we know that spotted cats have been around for at least 3,000 years, and of exactly the same shape and patterning as the Foreign-type spotted cats of today.

At the earliest cat shows, spotted tabbies of domestic type were exhibited and a contemporary writer described them as being plentiful, with spots of medium size and with sharp outlines, extending evenly over the entire body. By 1905 however, another writer reported that the spotted tabbies were becoming rare, and seldom seen at shows.

BRITISH SPOTTED TABBY
Neglected as a breed for many years, it was not until 1965 that the British Spotted appeared once more on the show bench; the old standards were revived and the variety recognized once more.

Today, the Spotted is a typical British Shorthair, with a stocky, hard and muscular body, strong legs

Brown Spotted Tabby British Shorthair – Grand Champion Brynbuboo Bosselot

with rounded paws and a fairly short, thick tail. Its head is very broad and massive with a short straight nose, firm chin, small rounded ears set wide apart, and large wide eyes. The coat is short and dense and covered with distinct spots. The head markings include the typical letter 'M' on the forehead and pencilled lines running from the corners of the eyes with more lines extending over the top of the head and down the neck.

The British Spotted is accepted in silver with black spots, brown with black spots, red with deep red spots and any other recognized ground colour with appropriately coloured spots.

The Silver Spotted must have green or hazel eyes, the Brown Spotted must have orange, hazel or deep yellow eyes, and the eyes of the Red Spotted should be deep copper.

There are no spotted longhaired varieties, possibly due to the fact that the length of coat would spoil the spotted effect.

ORIENTAL SPOTTED TABBY

In 1980, the Governing Council of the Cat Fancy approved championship status for spotted varieties of Foreign type. These had been originally bred in Britain to produce a breed approximating to the cat of the Ancient Egyptians, and called the Egyptian Mau.

Keeping in line with its policy of naming cats descriptively, the G.C.C.F. granted the title Oriental Spotted Tabby, and allowed a range of colours corresponding to those recognized originally for Tabby-pointed Siamese. Thus, fawn cats with black spots are known somewhat ambiguously as Brown Spotted Tabby, golden-beige cats with bronze spots are Chocolate Spotted Tabby, grey cats with slate coloured spots are Blue Spotted

Solitaire Kheta Saru, an elegant Chocolate Oriental Spotted

Tabby, and mushroom cats with lavender grey spots are Lilac Spotted Tabby. Red with Cream Spots is also allowed, but the beautiful silver variety with black spots is classed as 'experimental'.

Classic Tabby Cats

The Classic Tabby is also known as the Blotched Tabby or the Marbled Tabby in various parts of the world. Cats of this pattern must have markings which conform to an exacting standard. On the head there should be a letter 'M' on the forehead and a vertical line running from this over the back of the head to the shoulder markings. There should be an unbroken line running from the outer corner of each eye and pencilled markings on the cheeks.

On the neck and upper chest there should be unbroken necklaces of colour and the legs should be barred evenly with bracelets going right down to the toes. The tail should be similarly and evenly ringed. On the body, the two sides should be

Red Classic Tabby British Shorthair — Killinghall Red Jollity

identically marked, and should appear to have a butterfly drawn to overlay the shoulders. From the butterfly, a line runs down the spine-line to the root of the tail, and there should be a stripe on either side of this, running parallel to it.

These stripes are separated by lines of the ground colour. On each flank, the markings look like oyster shells, and are surrounded by one or more unbroken rings. The ground colour and markings should be evenly balanced and the pattern should have clear demarcation.

Classic Tabbies are recognized varieties in Long-hairs, and may be Brown, Red or Silver in Britain; a wider colour range is recognized in the USA. A similar range is accepted in the British and American Shorthairs, while in Orientals, Classic Tabby cats are classed as experimental.

BLACK CATS

As the natural or wild-type cat pattern is agouti, it is likely that all the first domesticated cats were tabby-patterned, being either ticked, striped or spotted. Then a spontaneous mutation must have occurred, causing the blotting out of patterning, and melanistic cats appeared, whose hair contained the black pigment melanin. Just like the melanistic jaguar, leopard, and some other varieties of wild cat, these were jet black in colour and would have mated freely with the wild-type cats until the black colouring was firmly established.

At the turn of this century, when the first cat shows were held, the Black Persian was popular. Black Shorthairs were also exhibited, and both breeds remain popular to this day. A recent addition to the designated black breeds is the Foreign Black, known as the Ebony in some countries. If you consider that 'Black is Beautiful', then whatever your preference for type, coat length or conformation, there is a black pedigree cat to suit your taste.

The Black Persian is a massive, stately animal with a short-coupled stocky body and a large round head. Its ears are tiny and set wide apart, and its huge round eyes of copper or deep orange glow like burning embers from the dense black face.

The British Black is similar in body structure to the Black Persian and has the same requirements of copper or deep orange eye colour. The coat is quite different, being dense, short and shiny. Scattered white hairs are considered a show fault, as are green rims to the brilliant eyes.

Known as the Ebony Oriental Shorthair in America and the Foreign Black in Britain – Champion Gangaili Schwarzer Teufel

The newcomer to the black trilogy was bred selectively from Siamese stock, producing a jet black, Siamese-shaped cat with clear green eyes. Long, lithe and sinuous, the Foreign Black retains most of its ancestors' characteristics, being talkative, lively and extrovert, with an affectionate nature.

For centuries black cats have had an air of mystique, being considered as good-luck charms in some areas, and harbingers of evil in others. A sailor would often keep a black cat at home, but preferred felines of different colours on board ship. Witches were said to favour black cats as their familiars, and many black magic rites and early medicinal remedies required the sacrifice of black cats.

BLUE CATS

Just as a simple mutation produced black cats from the original wild-type tabbies, so another natural mutation produced the dilution gene. Often referred to as the Maltese factor, this gene has the effect of reducing the intensity of colour in the cat's coat from black to blue. The possession of the factor changes a Black Persian to a Blue Persian, a British Black to a British Blue and a Foreign Black to a Foreign Blue.

Many breeds have a blue variety – there are Blue Balinese, Blue Colourpoints, Blue Burmese, Blue Rex, Blue Manx and many others, and some breeds are only recognized as blue such as the Russian Blue and the Korat. The Blue Persian or Longhair is the most popular of all the long-coated varieties and has been so for more than 80 years, since Queen Victoria owned a Blue and visited several cat shows.

The Blue Persian Society was formed in 1901 when the breed was a little lighter in bone than its modern counterpart, with a narrower head and larger pointed ears.

Today's Blue Longhair is a massive cat with a round, short head and tiny, wide-apart ears. The eyes are deep orange or copper in colour and complement the beautiful pale blue coat.

Any shade of blue is allowed in the official standard for the Blue Persian, and the breed is renowned for the length and fullness of the coat, the full short tail and the lovely long fringe of hair framing the face and known as the frill.

The British Blue is a powerful, chunky cat with similar conformation to that of its longhaired cousin, but a little longer in the foreface. Its full round eyes may be yellow, orange or copper in colour. The coat is short and dense, of a clear even blue colour, sound to the roots and without any signs of silvery tipping. It must be quite free from ghost tabby markings and speckled white hairs.

The Foreign Blue is a self-coloured blue cat of Siamese type and ancestry. It is an elegant cat with long svelte lines and the typical marten face of its ancestors, complemented by large, wide-set and wide-based ears. The coat is short, fine and close-lying and distinctly blue in colour, and the Oriental eyes are a clear green.

Blue cats have always been favoured and retained

as pets and have established themselves in many areas where there are restricted gene pools, such as small islands and remote country areas. Blue and white felines known as Cornish Cats (once bred in feral colonies in that county) and the earliest Cornish Rex cats, also carried the blue factor.

Blue cats are considered good luck charms in several countries of the world, and when cats were rare in the Western world, the blue specimens brought home by returning armies were among the most favoured.

The Korat

The Korat is a self-blue cat of modified Oriental or Foreign conformation, and is a natural, indigenous breed of Thailand, once called Siam. Named after a high and remote region in north-eastern Thailand

Blue Persian – Champion Evernden Andante

Korat queen Roro Si Suru Yothai of Si Sawat and her
delightful four-week-old kitten

known as the Korat Plateau, this rare and beautiful
cat has retained most of the characteristics of its
ancestor illustrated in the *Cat-Book Poems*, probably
composed in the thirteenth century. In this ancient
manuscript, carefully housed behind glass in the
National Library of Bangkok, several types of cat
are shown and described, and the painting of the
Korat is accompanied by the words: 'The cat Mal-ed
has a body colour like "Doklao" the hairs are
smooth with roots like clouds and tips like silver the
eyes shine like dewdrops on a lotus leaf'.

'Maled' means seed and refers to the seed of a
Thai plant called the Look Sawat, which has a
silvery grey fruit, slightly tinged with green; 'Dok'
means flower and 'lao' is a plant with silver-tipped
flowers. In Thailand, the Korat is called the Si-
Sawat, a compound word referring to its mingled
colours of silvered-grey and light green.

The Siamese King Chulalongkorn is said to have
named the Korat when he stopped to admire one of
the breed, and having enquired of its origin was told
that it came from the Korat area.

In its native land the Si-Sawat is rare and highly
prized, not only for its beauty, but because it is
considered to be a good luck charm. A pair of such
cats is a traditional wedding gift, symbolizing a gift
of silver, and given to bring prosperity to the couple
and to ensure a long and happy marriage. Sometimes such cats are given as a token of great esteem,
but they are rarely sold.

At the National Cat Club show in London in
1896, a Mr Spearman entered a Korat that he had
acquired in Siam. The judges disqualified the cat,
for they were of the opinion that any Siamese cat
should be fawn in colour and with dark brown
points!

In 1959 the first breeding pair of Korats arrived
in the United States, to be joined eventually by
other imports and in 1965 a special club was formed
to protect the breed and further its interest. The
breed was accepted by some American associations
in 1966 and by the rest by 1969. South Africa
recognized the breed in 1968 and Australia in 1969.

Korats first reached Britain in 1972, and met with
strong opposition from some cat fanciers who felt
that there were already sufficient blue breeds. By
1975, however, they had been officially approved by
the Governing Council of the Cat Fancy, although
they were denied championship status.

The Korat Cat Fanciers' Association, the breed

body in the USA, insists on its members abiding by a strict code of practice. All Korats must be able to trace their pedigree back to Thailand, and breeders pledge never to mate their cats to those of other breeds, never to deal with pet shops or animal dealers and to neuter all pet-type stock.

The Korat has a unique heart-shaped head with plenty of breadth between the eyes which are large and luminous, and seemingly too large for the face. When the eyes are closed they have an Oriental slant, and when they are open, they dazzle with their brilliance. The eye colour is green or may have an amber cast. The ears are large, with rounded tips and flared wide at the base. They are set rather high on the head, giving an alert, interested expression. There is a stop in the profile just in front of the eyes, and all the planes of the head curve gently and pleasingly with no sharp planes or angles.

The body of the Korat is neither as svelte as the Orientals nor as cobby as the Britishers, but it is hard and muscular, with an unexpected weightiness. The legs are in proportion to the body, and the tail is of medium length tapering to a rounded tip. The coat is single-layered and of short glossy hair lying close to the body. It is silver-blue in colour and as it catches the light appears to be tipped with silver. There are no tabby ghost markings anywhere, and never a sign of a stray white hair.

Russian Blue

One of the pedigree breeds which stands alone is the Russian Blue, for although it is a self-coloured blue cat of Foreign type, it has a standard of points ensuring that it does not resemble a straightforward Foreign Blue.

Russian Blue history is somewhat obscure, for in cat shows at the end of the last century it was common to lump all the Blue Shorthairs into one class, regardless of type. Russian fanciers were often disappointed when their sleek, fine-boned beauties were ousted from top placings by their chubby-cheeked British cousins.

At first, the Russian was known as the Archangel Blue, and the first cats of the breed were said to have been brought to Britain around the year 1860 by sailors returning from the Baltic port of Archangel.

Such cats were prized for their unique characteristics which included small triangular-shaped faces, clear blue coat colour and an unusual type of coat which, though extremely short, was soft and thick.

Frances Simpson, writing at the turn of the century, said:

The Blue Shorthaired cat, commonly called Russian, has a coat resembling plush in texture. These cats are supposed to have first come from Archangel, but the best authorities seem to agree in believing that they are not a distinct breed, and therefore they are now classed at our shows amongst the short-haired English varieties.

Blues should have deep orange eyes, and the colour of the coat may be light or dark, but must be even throughout, without any appearance of stripes or markings. A white spot, as in other self-coloured cats, is a blemish.

By the year 1912, the cat fanciers had come to an arrangement regarding their blue cats, and classes at shows were split to include those for Blue British-Type and Blue Foreign-Type, with the occasional indulgence of a Blue Russian-Type, for good measure.

Eventually, two cats were imported from Archangel by Mrs Carew-Cox, and did much to promote interest in the breed. Like some other varieties, the Russian Blue suffered a great set-back when breeders were forced to abandon their feline plans during the trying years of World War II, and all credit is due to those stalwarts who managed to keep their stock intact. As soon as hostilities ceased Miss Marie Rochford set about breeding up her now renowned Dunloe strain of Russian Blue cats.

Exports were made to the United States of America, where the breed soon became established, but then in Britain a retrograde step was taken. Following the Scandinavian example, Russian/Siamese crosses became commonplace. The Foreign Blue kittens born of these matings lacked the distinctive Russian coat, and when they matured and were mated *inter se*, often produced mediocre and unwanted Siamese-patterned offspring. It was not until 1965 that breeders made a serious attempt to re-establish the true Russian type

Russian Blue male – G.C.C.F. Champion
Czarist Nickolenka

and coat; a new standard was produced and careful breeding programmes were instituted using English and Swedish lines.

Even today, the true Russian coat is scarce on the show bench, and some breeders still find that long-forgotten recessive genes lurk within their prized cats' make-up, and the occasional Siamese-patterned kitten makes its unwelcome appearance among its true blue litter-mates.

Today's Russian Blue should have the special coat of short thick, double fur, which can be stroked both ways without exposing the blue skin. Judges prefer a medium blue colour, and when examined closely, some cats have silvered tips to each of the coat's hairs, giving an overall attractive sheen.

The Russian's type is a modified form of the Foreign Shorthair, for it should have long slender lines and small oval feet. The head is unique, being a short wedge in shape with a flat skull and a straight forehead and nose. The chin is firm and determined and the whisker pads should be very pronounced. The G.C.C.F. standard calls for large, wide-based ears set vertically to the head, while in America, the C.F.A. standard requires the ears to be set low down and far apart. Both standards agree that eye colour should be a vivid green, and that the eyes should be wide apart; in Britain they must be almond shaped, while the American standard specifies rounded eye apertures.

CHOCOLATE AND LILAC CATS

Chocolate and lilac are two colours rarely seen in pet cat populations, although these coat shades have been apparent in pedigree stock for some years.

Chocolate, like blue, is a modified form of the black pigment melanin. When the melanin is broken up and clumped together in one way the coat appears blue, and when the pigment clumps in a slightly different way, we see the coat colour as chocolate. Both these effects are caused by separate genes, and both chocolate and blue genes are recessive to black.

When a pure Chocolate cat is mated with a pure Blue cat, all the progeny are black because recessive genes must be inherited from both parents, before their effect becomes apparent in the offspring.

Lilac is an elusive colouring because it can only be produced when a chocolate cat receives a double dose of blue genes – both parents must have both chocolate and blue factors to achieve lilac kittens.

The first chocolate cats recorded were called Chocolate Siamese, and two types were imported from the East at the turn of the century, along with the Royal Cat of Siam – the Seal-pointed Siamese. The yellow-eyed and chocolate bodied 'Siamese' was obviously a Burmese or Burmese/Siamese hybrid and genetically black, while the blue-eyed Chocolate-pointed Siamese was not treated favourably, being considered inferior to the 'Royal'.

The Siamese Cat Club issued its official standard of points in 1902, emphasizing the 'Royal' and acknowledging the Chocolate, but the latter had to be registered as Seal-point to comply with the official form, and judges never placed them in the cards at shows. This resulted in most of the Chocolates being neutered and given away as pets, and it was only in 1950 that the G.C.C.F. finally gave them a separate breed number and allowed registrations to be amended. The Chocolate-point Cat Club was founded in 1954 and affiliated to the Council in 1961. In American too, Chocolate-point Siamese were recognized much later than Seal-point.

As many cats carrying the chocolate factor also carried blue, it was only natural that the ethereal Lilac-pointed Siamese turned up in occasional litters. At first these fairy-like kittens were considered to be poor Blues, but in America, the Lilac was recognized as a new variety – the Frost Point – in 1954. Five years later the conservative British Fancy also accepted it was a separate colour variety,

and the Lilac-point Siamese Cat Society was formed in 1961.

The Havana

In 1952 Isobel Munro-Smith wrote a cryptic note to the Siamese Cat Club. She was breeding Siamese with black shorthaired cats and hoping to achieve a truly black-pointed Siamese. In her letter she says:

The father of these Seal-Points is Tombee who has had some nice litters, among them some chocolate-points. Tombee and Susannah have produced the greatest thrill for all of us because, along with three blacks, they had a beautiful little *brown* kitten. It is male, Siamese in shape, with a long tail and nicely shaped eyes.

Mrs Munro-Smith then continues by asking about Burmese cats and wondering if this kitten could be a synthetic Burmese Brown. Born on 14 May, 1952, he was registered as Elmtower Bronze Idol, and became the first acknowledged Havana cat.

At the same time, another cat breeder was working on a study of the inheritance of the chocolate colouring in cats and planned to breed a

Chocolate Oriental Spotted Tabby kitten –
Solitaire Sekmet

self-chocolate cat of Foreign type. She was the Baroness von Ullman, and her preliminary matings were made between a Chocolate-pointed Siamese and a rangy black 'alley' cat.

Her friend, Mrs Hargreaves, mated her own Siamese and Russian Blue cats together, while another colleague, Mrs Fisher, started with Chocolate-pointed Siamese and half-Siamese Blacks. Together, this small group set out along the long path towards the creation of a new cat breed and on 2 June, 1953, the first planned Havana was born. He was Praha Gypka, and the pride and joy of Mrs Fisher's household.

Mrs Hargreaves' first Havana was Laurentide Brown Prior, born on 2 August, 1953 and Miss von Ullman's first was also a male, born early in 1954, and christened Roofspringer Peridot. These four males formed a healthy basis for the new breed and before long other enthusiastic breeders joined the group. By 1956, numbers had increased sufficiently for some exports to be made to the United States, where the cats were bred to different standards, eventually becoming a variety known as the Havana Brown, and quite unlike the modern Havana of Britain.

The self-chocolate Havana was named because it resembled the colour of a good Havana cigar, but when the Governing Council of the Cat Fancy granted recognition to the Chocolate Shorthairs in 1958, it insisted on the name Chestnut Brown Foreign, instead of Havana, presumably fearing confusion with the rabbit of the same name. This unwieldy and unpopular name persisted until 1970, when the Council relented and allowed a reversion to Havana.

During the 1960s, the breed lost favour when the continued breeding of like-to-like cats resulted in some loss of the desired Siamese type, and the development of dark coats and yellowish eyes. In 1970 fresh lines were started from good type Siamese stock. Havana cats were prominent among Best in Show winners and numbers soared.

The Foreign Lilac

As Siamese carrying blue as well as chocolate genes were used in the production of the Havana cat,

dove-grey kittens sometimes appeared in the litters and these were eventually developed as a variety in their own right, to be known as the Foreign Lilac.

Just as the Havana is the 'self' equivalent of the Chocolate-pointed Siamese, so the Foreign Lilac is the 'self' Lilac-pointed Siamese. The first of these exciting cats to be registered was Praha Allegro Agitato, born on 1 November, 1954, but due to lack of interest in him as a stud cat, he was neutered at an early age, and it was only in the late 1960s that a few breeders decided that the variety was worth developing.

Four separate lines of carefully selected stock were developed at the Harislau and Solitaire catteries, and eventually produced the third and fourth generations of like-to-like breeding necessary in those days to gain official recognition. Careful preliminary selection ensured good type from the outset of this breeding programme and the Foreign Lilac, originally bred as the Lavender Shorthair, was given full recognition in 1977.

Other Chocolate and Lilac Cats

Chocolate was introduced to the longhaired section of the Cat Fancy by two far-sighted breeders, Dr Manton and Mr Stirling-Webb.

Judicious crosses were made between Persians of good type and Chocolate-pointed Siamese, and generations of careful back-crossing were continued, with the object of developing the now famous and popular Colourpoints. Other lines were started using Chestnut Brown females and males, and in one early longhaired litter a self-chocolate longhaired kitten was born and named Briarry Bruno. Bruno was later matched with a chocolate-pointed Colourpoint, and as both parents also carried the blue factor, the resulting litter contained self-chocolate and self-lilac longhaired kittens.

Breeders of longhaired cats were slow to accept the potential of the Chocolate and Lilac varieties, possibly because of the work needed to attain really good type, but thanks to a small group, beautiful Chocolate and Lilac Longhairs may now be seen on the show bench and they attract a great deal of attention from the public.

Chocolate and Lilac varieties exist in other feline

An alert Lilac Oriental Spotted Tabby queen

groups too, including the Burmese, Rex and Abyssinian and the colours appear slightly different in each breed, being modified by other genes affecting the coat.

RED, CREAM AND TORTOISESHELL CATS

Red and Cream Cats

Red is the name given to the colour of pedigree cats which would be referred to as 'ginger' in the humble house cat. This rich bright coat is caused by the action of a sex-linked gene which follows simple rules of inheritance, even when modified by the dilution factor which changes red to cream.

Red and cream cats are found in all types of Fancy cats, and even in pet populations man has unconsciously selected for the colour, by picking out prettily marked ginger and tortoiseshell kittens to be retained, often allowing their black or tabbly litter mates to be destroyed.

There are Red Self and Red Tabby Longhairs, and both have been recognized since the earliest days of the Cat Fancy, when they were known as 'Orange, marked or non-marked.' Show catalogues of 1910 refer to the cats as red or orange, and by 1915, the colour had deepened sufficiently for the varieties to be called 'Red'. The dilute Red or Cream Longhair is first mentioned in feline literature by Harrison Weir in the late 1890s, writing of the early Angora cats imported into Britain. He called the Cream 'light fawn,' and one male of this colour named 'Cupid Bassanio' won many classes for his owner, despite the fact that his pale coat was heavily marked with tabby bars.

Breeders at the turn of the century understood little of colour inheritance in their cats, and the occasional appearance of a fawn kitten in an otherwise orange litter was looked upon as a sport or freak. Today, Red Self and Cream Longhairs should have clear, unmarked coats. The former is expected to be of a deep rich red, while the latter is a pale, clear buff shade.

In British Shorthairs, Self-red is classed as an experimental breed, but a few Red Tabbies are seen on the show bench and the Cream is very popular.

The Red Tabby Shorthair carries the same distinctive classic coat pattern as its longhaired cousin, etched in a deep red on a lighter red background. Due to the shorter hair, however, the pattern is more clearly defined, and therefore more striking. In American Shorthairs, both Red and Cream are recognized colours.

Very few shorthaired Creams are seen with totally unmarked coats, and the underlying tabby patterns of spotting, mackerel stripes or marbling are visible. Breeders of Burmese cats have had greater success in breeding their Red and Cream varieties, and these beautiful cats generally have sparkling clear and unmarked coats of deep guinea-gold and pale apricot, showing just the slightest ghost tabby as scarab markings between the ears.

In the Foreign Shorthairs, little success has been achieved with Self-red and Self-cream varieties, both of which usually show heavy ghost tabby markings, and are often mistakenly registered as Oriental Spotted or Classic Tabbies, despite the fact that they are genetically self-coloured. The Siamese group have experienced similar problems with Red-point and Cream-point Siamese, both of which are difficult to distinguish from their Tabby-point counterparts. The reason for this problem is the fact that the non-agouti gene which is responsible for producing self-coloured cats, as opposed to tabby ones, is quite ineffective on orange pigment. It is, therefore, only by careful selective breeding that unmarked reds and creams can be produced, and even then, the shorter and finer the coat, the more of the markings show through.

Perhaps the most interesting feature of the red colouring is in its sex-linked characteristic, which causes the production of Tortoiseshell females only, and gives rise to a great deal of confusion and misunderstanding among novice breeders.

Tortoiseshell

Tortoiseshell is the term applied to the cat's coat when it consists of a mixture of red and black hairs, generally clumped together to form random, clearly defined patches, but sometimes attractively intermingled.

In patched cats, the red or orange areas often

Red tabby-point Siamese

show ghost tabby markings, and this led to the erroneous belief that the tortoiseshell coat consisted of three colours, when genetically it must be said to have only two.

Dilute and recessive forms of tortoiseshell also occur in the cat. When the dilute gene is present, both pigments are reduced, the black becoming blue, and the red becoming cream, and this gives rise to beautiful Blue-Cream varieties. When the chocolate recessive gene affects the tortoiseshell pattern, the black is converted to chocolate brown, and the red to chocolate-red, producing rare Chocolate-Torties or Chocolate-Creams.

Only when both dilute and chocolate genes are present is it possible to produce the Lilac-Tortie or Lilac-Cream varieties, while the black is reduced to pale lilac, and the red becomes a light pastel beige in tone.

Tortoiseshell cats are usually female, and any males born with the coat pattern are generally sterile, although there have been a few exceptions to this rule, for the coat pattern in Tortoiseshell varieties is produced by the action of a semi-sex-linked gene, as we shall see.

There are Tortoiseshell and Tortie-and-White Longhairs or Persians, and Tortoiseshell-pointed Colourpoints or Himalayans. Blue-Cream Longhairs are an old-established breed and the dilute Calico or Blue-Cream-and-White is being revived. Colourpoint breeders have produced some beautiful chocolate and lilac-tortie cats with colour restricted to their points, as well as Chocolate-Tortie and Lilac-Tortie Longhairs.

In the British Shorthairs, Torties and Tortie-and-Whites have been popular for many years, and their dilute varieties, the Blue-Cream and Blue-Tortie-

and-White, have also been bred. Now that Chocolate British Shorthairs are appearing, it is to be expected that Chocolate and Lilac Tortoiseshells will be bred in the future.

Burmese are accepted with tortoiseshell coats in all four basic colours, and Siamese are accepted with Seal-Tortie, Blue-Tortie, Chocolate-Tortie and Lilac-Tortie points.

In the Oriental Shorthairs Tortoiseshells of all colours are being bred, but at present may only be shown in Assessment classes. As Rex cats are allowed in all accepted feline coat colours and patterns, many Tortoiseshell-and-White specimens are seen on the show bench and are particularly attractive.

The appeal of the tortoiseshell coat seems to be in its often bizarre distribution, and the presence of a coloured blaze bisecting the face from forehead to nose often adds to the harlequin effect.

Each tortie is uniquely patterned – no two are ever alike – an attraction to those who like to have an unusual pet. Conversely some people find the random patterning displeasing to the eye.

Dilute Tortoiseshell or Blue-Cream

Towards the end of the nineteenth century, two of the most popular Persians were the Blue and the Fawn, now known as the Cream. Some breeders mated the two colours together and were surprised, and somewhat alarmed, to find that all the female offspring from such matings were mottled, instead of being self-coloured like their parents. These females were referred to as 'marked blues' and were usually given away as pets, while their brothers, always resembling their dam in colouring, were considered suitable for showing and breeding.

In 1901, one breeder decided to keep a particularly good 'marked blue' and mated her to a Fawn male. To her amazement and delight, of the four female kittens in the litter, only two were 'marked blues' and the other two were Fawn – exactly like their father. Despite this discovery, many years passed before the Cat Fancy deduced that a sex-linked gene was at work in this variety. Lots of people noticed the 'marked blues', and when the Fawn Persian was eventually re-christened the Cream, the 'marked blue' was named the Blue-Cream.

Eventually, the G.C.C.F. granted recognition to the breed in 1930, despite the fact that only females had so far been born. Even as late as 1939, one breeder wrote that she was hoping to mate Blue-Cream to Blue-Cream as soon as she had produced some males.

The war years forced many breeders to abandon their plans, but eventually good quality kittens appeared once more and it was realized, at long last, that the Blue-Cream is an all-female variety. Males are very rare indeed and are generally found to be sterile. Now that more is known about the genetics of coat colour in the cat, and the sex-linked character of the orange gene is understood, mating results can be accurately predicted.

In type and conformation, today's Blue-Cream Persian is generally as good as the Blue and the Cream, and the softly intermingled coat colouring, called for in the G.C.C.F. standard, has produced an ethereal-looking cat with a profuse coat of pastel blue and palest cream hairs.

Although the standard does not mention a cream blaze bisecting the Blue-Cream's face, many breeders like their kittens to have this feature and it certainly adds to the variety's appealing expression. The eyes must be deep orange or copper, and like those of all Persians, are large and lustrous. The Blue-Cream's official G.C.C.F. Breed Number is 13, but it has been far from unlucky and is present in large numbers on the show benches today.

The British Blue-Cream is similar in colouring to the Longhair variety, and its G.C.C.F. breed number is 28. The first Blue-Cream Shorthairs began to emerge about 1950, when, due to a shortage of stock following World War II, breeders were forced to mate together cats of different colours. Shown in Any Other Variety classes at cat shows, the Blue-Cream soon captured the hearts of several dedicated cat fanciers who worked the variety up to recognition standard by 1956.

Good type is apparent in the beautiful Blue-Cream Shorthairs seen at today's shows. Their softly intermingled blue-cream coats are complemented by large round eyes of yellow, orange or copper.

It is interesting to note that the C.F.A. standard of points for Blue-Cream cats in America calls for distinct blue and cream patching, with no intermingling of colours; in Britain, such cats would be heavily penalized.

The Orange Gene

The gene responsible for red pigmentation is referred to as orange, and is generally symbolized by a capital O while its opposite form, non-orange, is symbolized by a lower case o.

The sex of a kitten is determined by a special pair of chromosomes, the X and the Y. A female cat has the make-up XX and therefore imparts X chromosomes to her offspring. The male cat has an XY make-up, and therefore can impart either an X or a Y to his offspring.

If a kitten receives an X from its sire as well as the one from its dam, it will be female – XX. If it receives a Y from its sire, it will then be an XY, and therefore a male.

The O gene can only be transmitted along with the X chromosome in cats, and this means that the orange male can only transmit the gene to his female offspring.

A simple table is given here, showing the results of some Orange matings. Remember, orange means red or 'ginger' of all sorts, and the rules hold true whether the cats are Longhairs, Shorthairs, British, American, Foreign, Burmese or Siamese, and with or without white. When the dilution factors are present on both sides, Cream can be substituted for orange, and Blue-Cream for Tortoiseshell. Where the term 'black' is used, it really denotes 'non-orange' and can be any non-orange colour, such as black, blue, chocolate and so on.

A simple way to remember this is to learn the basic rules which state that both parents must show orange to produce orange female kittens – the sire must be orange and the dam either orange or Tortie; orange parents pass their colour genes to kittens of the opposite sex; orange cannot be carried without showing – but may be present in only a few hairs in the cat's coat; 'orange' rules are applicable even when the colour manifests itself in its dilute and recessive forms.

Expectations of kittens from matings involving the Orange (sex-linked) gene:

Parents		Kitten	
Male	**Female**	**Male**	**Female**
Orange	Orange	Orange	Orange
Orange	Black	Black	Tortoiseshell
Orange	Tortoiseshell	Orange	Orange
		Black	Tortoiseshell
Black	Tortoiseshell	Orange	Tortoiseshell
		Black	Black
Black	Orange	Orange	Tortoiseshell
Black	Black	Black	Black

From these simple rules we may see that a tortoiseshell female passes her orange colour or her black colour to her male offspring, whatever the colour of her mate, while an orange male passes his colour to his female offspring, but unless the queen he mates is also orange, or tortie, the female kittens will be Tortoiseshell, not completely orange.

Some Tortoiseshell cats have patches of random white spotting present in their coat patterns, and are then said to be Tortoiseshell-and-White, or Calico in the USA. The same rules for inheritance of orange apply to these varieties, and they also occur in dilute and recessive forms.

The tortoiseshell coat pattern is present in almost all of the pedigree varieties of domestic cat as well as being common in the mongrel and feral cat populations.

WHITE CATS

The recognized varieties of white cats in the Cat Fancy, regardless of their conformation or coat length, owe their colour, or non-colour, to the effect of the dominant white gene, referred to by geneticists as W. This manifests itself by masking the cat's true colour genotype, and is capable of covering any other colour or pattern. This means that most white cats will produce variously coloured offspring, depending on the genotype of their mates.

If a cat is homozygous for white, with a genotype of WW, then all its offspring will be white. However, a great many white cats are heterozygous for white with a genotype of Ww and will produce some coloured kittens unless mated to a homozygous white mate.

Most white cats occur in three sub-varieties, showing differences in eye-colour, and may be orange-eyed (known as copper-eyed in America), blue-eyed or odd-eyed, in which one eye is orange and one eye is blue. This holds true for the White Longhair or White Persian, and the British or American White Shorthair. The exception is the Foreign White which inherits specially pigmented blue eyes from its Siamese ancestors.

The white gene sometimes causes deafness, particularly in the blue-eyed Longhair and Shorthair varieties, by affecting the cochlea shortly after its birth. This deafness, which is irreversible, rarely troubles the white cat kept confined as a house pet, but can seriously threaten the safety of the white cat allowed to roam free, unable to hear approaching dangers such as traffic or predators.

Deaf white queens, unable to hear their kittens' cries, often appear indifferent to the litter's needs and the youngsters may prove difficult to rear successfully. Breeders are advised not to use deaf individuals in their programmes and, in this way, the incidence of deafness in white cats could be significantly reduced.

White Longhairs

In Longhairs, the Blue-eyed White has the longest history. Specimens of the breed were quite numerous at early cat shows, and when breed registrations first began, had their own number, championship status and separate show classes.

In the late 1930s problems with deafness had been encountered and it was decided that judicious out-crosses to top quality Blue Persians would counteract the defect, and also improve overall body type. The resulting white kittens had blue eyes, orange eyes and odd eyes and eventually the G.C.C.F. decided to accept the Blue-eyed White and the Orange-eyed White as separate varieties, while the

Odd-eyed Whites were considered to be White Longhairs with incorrect eye-colour.

Today, all three sub-varieties enjoy full show status and conform to the ideal standards for pedigree Persian cats. These Whites have very long, profuse coats which repay correct daily grooming, and the judicious use of talcum powder helps to clean the fur and keep it from matting. Before shows, White Longhairs benefit from gentle bathing, and when properly prepared are among the most glamorous and eye-catching of all show stars.

White Shorthairs

White Shorthairs were popular at the turn of the century and commanded relatively high prices among fanciers. As in the White Longhair, the Blue-eyed variety was found occasionally to exhibit some degree of deafness, and out-crosses were made to produce an Orange-eyed variety which, in turn, gave rise to Odd-eyed Whites. Today, all sub-varieties are accepted for championship competitions.

White cats may also result from the effect of a gene c^a, which removes all pigment from the coat, producing an albino variety with pale blue eye-colour. This variety has been developed to some degree in the United States as the American Albino Siamese, considered to be a Siamese cat without pigment at the points. As c^a is a recessive gene, this 'whiteness' can be carried, but cats with the gene have been eliminated from British fanciers' breeding programmes as it was considered unwise to perpetuate the albino as a pedigree variety.

Foreign Whites

The first steps in developing the Foreign White as a separate variety were taken in 1962, when three independent breeders in Britain decided to cross Siamese cats with ordinary White Shorthairs. They all had the same intention, although quite unknown to one another, and this was to breed a pure white cat of Siamese type and conformation, and without the problem of deafness so often encountered in existing white cat breeds.

By 1966 the interested breeders and their friends

had joined forces and formed a specialist breed society. Expert genetic advice was sought and strict breeding plans were formulated. At each generation, the shorthaired white cats were mated back to the top-quality Siamese and eventually the desired type and intensity of eye-colour was produced. In 1977 the breed was granted full recognition and champsionship status by the Governing Council of the Cat Fancy and within two years the first Grand Champion of the variety was proclaimed.

Today's Foreign White exactly resembles the Siamese in type, and has deep sapphire-blue eyes. It is a cat of medium size with long slender limbs and neat oval paws; the neck and body are long and elegant and the tail is slender and tapered. The striking head has straight planes forming a distinct wedge shape, accentuated by the large, wide-apart and wide-based ears. Just as in the official Siamese standard, the Foreign White is faulted if it has a squint or a kink in the tail.

BI-COLOURED CATS

The term Bi-coloured is generally applied to cats of recognized fancy feline colours with the addition of large areas of white patching. This effect is caused by the presence of a factor known as the white spotting gene, and its effect is to mask off portions of the coloured coat with white areas.

By careful selective breeding it has been possible to produce exhibition standard Bi-coloured cats which are evenly marked, rather than the randomly-patterned cats often seen in mongrel populations. There are Bi-coloured varieties in both Longhairs and British and American Shorthairs, and apart from the coat length both have similar standards of points.

Pedigree Bi-coloured cats may be of any solid colour and white and the coloured areas must be clear and evenly distributed, symmetry of design being highly desirable. Not more than two-thirds of the coat should be coloured and not more than one-half of the coat should be white. The face should be patched with colour, preferably bisected by a white blaze and the large lustrous eyes should be of deep orange or copper.

A Japanese Bobtail kitten with 'Mi-ke' pattern plays happily at Swady's Cattery in Seattle

Show cats in these varieties are penalized for having small patches of white within the coloured areas of the coat, and for any brindled or tabby markings – it is certainly an exacting task to breed a top show Bi-colour.

Piebald cats have been beloved for many years and were common in the earliest cat shows. In 1889, it is recorded, several shorthaired, bi-coloured cats were exhibited. There were Black-and-White and White-and-Black, plus Brown, Red, Blue and Yellow Tabbies, all with white. In those days obviously, tabbies were allowed. The standards then called for white areas to be present on the nose, chest and feet.

Cream and White Bi-colour Persian male –
Thomas Tickerty Boo

At the turn of the century a well-known cat fancier, Lady Alexander, tried to popularize the Bi-colour by guaranteeing special classes at the London shows, and although numbers were small, great interest was aroused in the Black-and-White Bi-Colours which were known as Magpie cats.

Eventually, the Governing Council in Britain approved exacting standards for the Bi-colours in which the desired colour distribution was to be exactly as found in the Dutch rabbit. Few cats were able to meet the requirements and many breeders lost heart, deciding to breed other, less demanding breeds. In 1971 a revised standard was approved, and since then the Bi-colours have made their comeback, and many have achieved top-show awards.

SILVER CATS

Silver colouring in cats is caused by the effect of a gene known as the Inhibitor, which is symbolized as I by geneticists. It works by suppressing the development of pigment, or true colour, in parts of the cat's coat and has a greater effect in the lighter and generally less pigmented areas.

In a tabby cat, the lightest areas are those of the agouti base coat as opposed to the solid coloured areas which produce the characteristic tabby markings.

When the Inhibitor is added to a British Classic Tabby, for instance, the result is a cat with a striking black marbled pattern clearly etched against an almost white base coat, producing the cat known as the British Silver Tabby. There are Silver Tabby Longhairs too, and although the pattern is somewhat diffused by the long coat, they are an especially beautiful breed. Silver Spotteds are another variety of the British Tabby, and in these, the light base coat is clearly dotted with randomly placed black spots, producing a cat that resembles a miniature snow leopard.

Oriental Shorthairs have been developed in both classic patterned and spotted Silver Oriental Tabbies, but these are still in the experimental stage as far as the G.C.C.F. is concerned, although they are recognized as varieties of the Oriental Shorthair in the USA. With their long, elegant lines and sharply defined patterning on silver base coats, these cats are eagerly awaited on the show scene.

Breeders of Oriental cats have introduced the Inhibitor gene to cats of all colours, and while many beautiful kittens have been produced it has been difficult to determine, in many cases, whether the offspring are silvered or self, and Oriental Smoke kittens can be particularly difficult to assess.

When 'silver' is present with colours other than black, the reduction of the base coat pigment is affected and the results may be far from those expected, so dabbling with the Inhibitor gene should not be attempted by the novice breeder. Breeding programmes need to be carefully formulated and meticulously followed in the production of silvered cats. Experience is vital in kitten assessment too, for in most silvered varieties, young kittens often exhibit markings totally different to the ones they will possess when they reach maturity.

Sometimes it is easy to determine Smoke kittens at the moment of birth, before their fur dries out,

for the silvering may be clearly visible. Once they have dried, they appear self-coloured and the silvering may not reappear for many weeks. Some kittens thought to be Smoke may be merely self-coloured, with very unsound coats, while true Smokes may appear to be self-coloured and often show some ghost tabby markings too.

Apart from these pitfalls, adding Inhibitor to existing varieties is comparatively simple. The gene, being dominant to full colour, only needs to be present in one parent in order to produce silvered offspring, and its dominant character also ensures that it cannot be carried invisibly – non-silvered offspring of a silvered parent cannot pass on the Inhibitor gene.

Abyssinian cats have been bred in a silver variety for several years. The Silver Abyssinian is a beautiful variety, each silver hair banded with colour to give a uniquely ticked and sparkling effect. This breed too, is in its experimental phase.

Perhaps the most popular of all silvered cats is the photogenic Chinchilla, a longhaired breed in which the Inhibitor gene affects virtually all of the coat,

Silver Tabby American Shorthair – Grand Champion Portrait's Romeow

producing a white cat, coloured only at the very tips of each hair, giving rise to the variety's spangled appearance.

Between the two extremes of the lightly marked Chinchilla and the Smoke Persian, a cat so densely coloured that it appears almost black apart from its frill and ear tufts, are two varieties known as the Shaded Silver and the Pewter. These are virtually identical, being longhaired silver-white cats whose coats are overlaid with pewter-coloured mantles. The Shaded Silver, however, has green eye colour, while that of the Pewter should be orange or copper.

The Cameo series of longhaired cats is beautiful, and resulted from the marriage of sex-linked orange and Inhibitor genes. The range of coat colours is equivalent to that seen in the Chinchilla, Shaded Silver and Smoke.

The Smoke Shorthair, like the Smoke Persian, has been bred for some years. Its coat is heavily pigmented from the tips of the hairs to about half-way down their length, and the Inhibitor-affected base coat shows through particularly around the head and the underbody.

British Tipped

British Shorthair breeders have successfully developed short-coated versions of the Chinchilla and the Cameo Persians too, and the variety known as the British Tipped has achieved rapid acclaim.

The Tipped may have its coloured areas in any of the recognized British Shorthair colours, plus chocolate and lilac. The original and most striking of the Tipped cats is a white cat with black tipping and green eye colour. The other Tipped cats' colours lack the brilliant contrast but are nevertheless very attractive, and must have orange or copper eyes.

Chinchilla

Today's exhibition Chinchilla must surely rank among the most beautiful of all fancy felines with its black tipped and sparkling white coat, and its amazing, expressive eyes. The sparkling quality of this unique cat is produced by an extreme effect of the dominant Inhibitor gene which suppresses the

development of colour, in this instance in all but the tips of each of the coat hairs.

The Chinchilla is an old-established breed and has always enjoyed popularity. Queen Victoria's grand-daughter Princess Victoria took her cat-breeding seriously and was patron of the National Cat Club for some time. She specialized in Chinchilla cats as well as Blue Persians, and owned a stud male called Puck III. Old records refer to a cat owned by a Mrs Vallence. This was Chinnie, the product of a pedigree queen and a roving tom cat. Said to have been of 'a silver tone . . . really a light lavender', she was mated to a famous stud called Fluffy I, and later, their kitten Beauty matured and was mated to Champion Perso, a light Smoke male. Their son Silver Lambkin became the first true Chinchilla stud cat and his appearance at the Crystal Palace Show in 1888 caused a stir when he succeeded in sweeping the board. When this remarkable cat eventually died, aged 17, his body was preserved by the taxidermist, and exhibited in the Natural History Museum in South Kensington.

In 1894, Chinchilla cats were given a class of their own for the first time. They were rather different to the show Chinchillas of today, having been produced from matings between longhaired Silver Tabbies and Blue Persians. They had light undercoats with mantels of slate grey. It is this distinctive colouring that is thought to have given the breed its name, for the original cats closely resembled the attractive little fur-bearing rodent *Chinchilla laniger*.

The Chinchilla Cat Club was formed in 1901 and its official standard of points for the Chinchilla reads: 'Silvers or Chinchillas should be as pale and unmarked a silver as it is possible to breed them. Any brown or cream tinge a great drawback. The eyes to be green or orange.'

Today's G.C.C.F. standard is far more explicit (the American standard is very similar), calling for a pure white undercoat and black tipping evenly distributed over the coat on the back, flanks, head, ears and tail, giving the breed its characteristic sparkling silver appearance. The legs are only slightly shaded and the chin, stomach, chest and ear tufts must be white. Brownish or creamy markings are disallowed and any sign of tabby patterning is penalized. The coat must be of fine texture, long, dense and silky and with a distinctive frill of even longer fur around the neck, framing the face. The Chinchilla should have a cobby body and short, sturdy legs. The head must be round with plenty of breadth between the tiny tufted ears. The snub nose is brick red in colour, and the large, round and expressive eyes may be either emerald green or acquamarine in colour, the lids rimmed with black or dark brown pigment, adding to their lustre.

Shaded Silver and Pewter

As the years went by and cat breeding became more popular, it was noticed that the Chinchillas were produced with either lightly ticked or heavily ticked coats. In 1901 they were divided into the Silver or Chinchilla as one variety, and the Shaded Silver as another, despite the fact that kittens of both types were to be found in the same litters.

In Frances Simpson's *Cats for Pleasure or Profit*, 1905, the contemporary standards are detailed:

Silvers or Chinchillas should be as pale and unmarked a silver as it is possible to breed them. Any brown or cream tinge a great drawback. The eyes to be green or orange. . . . Shaded Silvers should be a pale clear silver, shaded on face, legs and back, but having as few tabby markings as possible; eyes green or orange. Any cream or brown tinge a great drawback.

Unfortunately for their British fans, the Shaded Silver standard was later withdrawn, and only Chinchillas were recognized for competition, although the Shaded cats were still accepted in Australia, New Zealand and the United States of America. In more recent times, the Shaded Silver made a comeback in Britain and although it had to compete in Any Other Variety classes, under the blanket breed number of 13a, cats of this variety made quite an impact on felinophiles and provided an important contribution to the breeding of the Cameo series of Longhairs, which result from the combination of red and silver genes. Breeders began to select for either green or orange eye colour, and decided that they wanted the cats with green eyes to be designated Shaded Silver, and those with orange eyes to be called Pewter.

In 1978 the Pewter Persian suffered a great setback in Britain, when it lost its 13a breed number and was relegated to experimental status, having to appear in assessment classes, the cats being judged only against their standard of points instead of being in competition with one another.

Eventually, in 1980, the variety was given its own breed number, enabling it to compete on the show bench, but to date, championship status has not been granted. Today's perfect Pewter is a truly beautiful cat, white in colour, evenly shaded with black, which gives the effect that the animal wears a pewter-coloured mantle. The ear tufts, chin and underparts are pure white, while the legs and tail are shaded. The silky coat is long and dense, with a full frill of fur framing the face. The round head has a snub nose and plenty of width between the neat ears. The chin is firm and the teeth are level. The Pewter's expressive eyes are large and round, and the orange eye colour is enhanced by a rim of black pigmentation around the eyelids. Show Pewters are penalized for having green rims to their eyes, kinks in their tails or heavy tabby markings. Any sign of cream or brown tarnishing is also considered a fault.

Smoke Longhairs

Smoke colouring in cats is caused by the effect of the Inhibitor gene on a normally self-coloured feline coat, producing coloured cats with near-white undercoats, which often appear self-coloured in repose. The Smoke Longhair is an old-established breed, and one was recorded by Harrison Weir who visited a cat show in 1872: '. . . a beauty was shown at Brighton which was white with black tips to the hair, the white being scarcely visible unless the hair was parted . . .'. In 1893 the Smokes had their own show classes, and by 1900, the Silver and Smoke Persian Cat Society was formed.

Often called 'the cat of contrasts' the Smoke should be of outstanding Persian conformation, its silver ruff or frill framing a jet black face enhanced by full round eyes of deep orange or copper. The neat black ears should sport silver tufts, and the dense body coat is black, shading to silver-white on the flanks and sides.

The Blue Smoke is the dilute variety, very attractive, but less dramatic in effect than the Black Smoke.

Champion Oxus Tarquinius Superbus – an aptly named Black Smoke

The Cameo is a comparatively new variety and the darkest of the Cameo group. With its rich red coat with silver shimmering through, and its deep copper eye colour, a good specimen always causes a stir at shows.

BRITISH SMOKE

In the British Shorthair group, Smoke varieties are accepted in both Black and Blue. The Standard calls for the cats to be compact and powerfully built, with short strong legs and a tail thick at the base. The head of the British Smoke should be massive and round, set on a short thick neck. The small rounded ears should be set wide apart and so should the full round eyes of yellow or orange. Show cats of these varieties are penalized for having white guard hairs, incorrect eye colour, over-long coats or tabby markings.

ORIENTAL SMOKE

Oriental Smokes are bred in a wide range of colours but are, as yet, unrecognized varieties. They all have true Foreign type, being similar in conformation to the show-standard Siamese and are svelte and elegant cats. The head of the Oriental Smoke should be long and wedge-shaped with straight planes from all angles – no whisker break and no dip in the profile. The large ears are wide at the bases and set well apart, continuing the lines of the cheeks and accentuating the wedge of the head. The expressive eyes should be almond-shaped and slanted towards the nose.

The Black Oriental Smoke looks similar to a Foreign Black, but the silver underlay is visible around the face, the underparts and often between the legs.

The Chocolate Oriental Smoke is similar to the Havana, and again the light undercoat shows most clearly on the head and underparts. There are Blue, Lilac and other colours in Oriental Smokes too, and a small group of breeders are working towards their acceptance.

Cameo Cats

The Cameo was first bred in the United States by Dr Rachel Salisbury in 1954, using Smoke and Tortoiseshell Persians for the initial crosses in order to combine the necessary Inhibitor and orange genes which produce the Cameo effect. Some breeders used Chinchilla stock to introduce the Inhibitor gene but this also had the effect of introducing green to the eye-colour, an undesirable trait, and the practice was quickly discontinued.

Eventually, in 1960, Cat Fanciers' Association granted the breed full recognition in the United States and Cameo cats were soon taking top show honours.

In Britain, a few dedicated breeders set off along the long road to breeding the perfect Cameo, and despite many setbacks, achieved Governing Council acceptance and an official breed number in 1982, allowing Shell, Shaded and Smoke varieties in both red and cream. The official standard gives a large percentage of the points for correct coat colour and degree of tipping, and also puts emphasis on correct head type. Any sign of barring or tabby markings is considered a serious show defect.

Cameo kittens are born light, developing their coloured tipping as they grow. They prove to be forward and affectionate kittens. Cameo cats are totally captivating and always cause great interest among visitors to shows at which they are exhibited. They are chunky cats of typical longhaired type and conformation, with short, strong legs. The Cameo's head should be broad and round with a snub nose and good width across the head between the tiny tufted ears. The glorious large round eyes should be deep orange to copper in tone, and without any sign of green to the rim, which is considered to be a serious show fault.

The long coat must be soft and flowing. The Cameo is a cat of contrasts, the undercoat being almost white, and each hair of the topcoat being tipped with colour. The deepest intensity of colour is seen on the face, the legs and feet along the spine-line from the nape of the neck to the tail-tip. The lightest areas are the frill of long fur around the neck, the underparts and the long tufts in the ears.

There are three varieties of red Cameos, named to correspond with the degree of tipping present and thus with the overall effect of the colour intensity observed. The palest of all is the Red Shell in which the near-white coat is just tipped with light red, giving a sparkling, pinkish effect.

The tipping of the Red Shaded Cameo extends down a greater portion of each hair tip, giving the cat the effect of wearing a red overmantle.

The Red Smoke Cameo appears just like a self-coloured red cat until it moves, when the rippling motion of the coat allows the white undercoat to shine through. In the United States, the three shades are sometimes known as the Red Chinchilla, the Red Shaded and the Cameo Red.

Much paler than the red series of Cameo cats is the cream group in which the red tipping is replaced by cream, due to the effect of the dilution gene.

The Cream Shell Cameo is an ethereal-looking cat of palest peachy-pink, while its shaded counterpart is a little more intensely coloured. The Cream Smoke also appears to be a self-coloured cat while in repose. Female Cameos are also found as Tortoiseshell and Blue-Cream varieties. The former has tipping comprising areas of black and red, well patched over the body and face, and without any solid areas of colour on the feet and legs. The latter has tipping of softly intermingled shades of blue and cream.

4

Colours, Patterns and Breeds II

THE HIMALAYAN FACTOR

The Himalayan factor, or Siamese gene, is responsible for reducing the colour of a cat to the extremities of its body, known as its 'points'. The gene is one in the albino series in the cat, and is related to those which produce Burmese and total albinism. It acts in a recessive manner to full coat colour, and only when both parents have the gene can Himalayan-patterned kittens be born. The distinctive 'pointed' appearance of Himalayan-patterned cats is due to partial de-pigmentation of the coat over the warmer areas of the body, and the gene that causes this is also responsible for partial de-pigmentation in the eye, causing the cats to have blue eye-colour.

The intensity of the de-pigmentation is thermally affected to a certain extent, and Himalayan-patterned cats reared in cool conditions generally have darker coats than those kept very warm. Operations and accidents which affect body temperature in discrete areas of the skin affect the new hair growth, and a Siamese female which has a small area of her flank shaved for the spaying operation will grow a patch of darker fur in that region. This grows out with the next hair growth, usually at the spring or autumn moult.

Siamese

The first known of all Himalayan-patterned cats was the now familiar Siamese, so christened because the

Seal-point Siamese stud male Champion Shandean Glastron tests the upholstery

original imports to Britain came from Thailand, then called Siam. Old Thai manuscripts showing various coat patterns in domestic cats clearly depict the Siamese with a white coat and black extremities and the earliest reference to the breed is seen in the thirteenth-century *Cat-Book* Poems, the manuscript of which may be seen in the National Library of Bangkok.

The relevant text reads:

> The eighth picture is of a cat called
> WI-CHI-AN-MAAD (Diamond)
> The Upper Part of the Mouth
> The Tail, All Four Feet and the
> Ears
> These Eight Places
> Are Black
> The Eyes
> Are Reddish-gold in Colour
> The Cat Called Wichianmaad
> Has White Fur

Of course we know that Siamese in fact have blue eyes, but these glow reddish-gold at night with reflected light. While this proves that such cats existed in Siam, hundreds of years ago, it does not necessarily mean the gene mutated in that country.

Peter Simon Pallas, the great German-born naturalist, saw a Siamese cat during his exploration of the Caspian Sea towards the end of the eighteenth century. He described it as having a light chestnut-brown body, paler along the sides and underbelly and black on the face, ears, paws and tail. He did not mention the eye colour, but says the head was longer towards the nose and the cat generally smaller than a common cat.

Siamese cats were first introduced into England and to the United States of America in the latter part of the nineteenth century and caused a sensation when exhibited at the Crystal Palace Show in London, 1871. One report described the Siamese as '. . . singular and elegant in their smooth skins, and ears tipped with black, and blue eyes with red pupils', while another called the breed 'an unnatural, nightmare kind of cat'.

Early imports were chunky, apple-headed cats and although light-coated as kittens, soon darkened with age. Many older Siamese fanciers of today deplore modern dark-coated Siamese and extol the pale-coated cats of yesteryear. However, Frances Simpson's works of the early 1900s lead us to believe that the natural darkening process has always been apparent. She says '. . . it is a pity that Siamese cats gradually lose the beautiful pale fawn colour and their coats darken as they grow older. It is quite the exception to see a grown-up Siamese light in body colour . . .'.

There is little doubt that the original Siamese were all genetically black, the forerunners of the type today which has dark brown points and is called Seal-pointed, or Royal Cat of Siam. Some of these carried recessive and dilute genes so that in following years Blue-point and Chocolate-point cats naturally occurred, although it took the Governing Body many years officially to recognize them. It is possible that these different colours were treated as 'sports' or freaks in the early days, and as Siamese cats were often considered as good-luck charms, such oddities might well have been destroyed as harbingers of evil. Later, ethereal, pale-pointed kittens began to appear, but were usually discarded as being 'poor' Blue-points. As the science of genetics was applied to coat coloration in cats, however, it became apparent that these kittens were Lilac-points, produced only when both parents carried both chocolate and blue genes.

Not content with four colours in their Siamese, some breeders decided to introduce the sex-linked red colour, and although it took several years of selective breeding, back-crossing and effort to gain approval, Red-point, Cream-point and four shades of Tortoiseshell-point were eventually given G.C.C.F. approval.

Other Siamese, in making their own mating arrangements, had produced the occasional tabby, or even tabby-pointed offspring, and some Siamese fanciers thought that these were so attractive that they should be developed as a separate variety. First called Shadow-points, then Lynx-points, a name still favoured by some American Associations, the Tabby-point Siamese was gradually bred to exhibition standard. It is now seen in the four standard shades of Seal, Blue, Chocolate and Lilac, broken into tabby patterns at the points. All Siamese should retain pale, clear body colour, and all have blue eyes, the colour intensity present depending upon the density of the point colour. Type is constant whatever the colour, and is far removed from that of the little cats which originally came from Thailand.

Red Tabby-points and Cream Tabby-points are difficult to distinguish from 'self' Red and Cream-points, for all show similar shadowy tabby markings. There are Tortie-Tabby-pointed females too,

Cream-point Siamese with a splendid set of whiskers –
Palantir Tom Bombadil

in all four standard shades, but although these can be useful for breeding, they are not considered to be particularly attractive cats.

Today's Siamese is long and svelte, every inch a thoroughbred, with fine bone and elegant lines. The long wedge-shaped head is accentuated by wide-set ears and oriental eyes, and the coat should be very short, fine and close-lying. Some of the original Siamese had shortened, twisted or kinked tails, and occasionally squinted with one or both eyes, but the modern Siamese is heavily penalized for such defects at shows.

Balinese

The lithe and graceful Balinese was named after the renowned dancers of Bali. It is, in fact, merely a long-coated Siamese cat and has been bred from the occasional long-coated kittens which appeared from time to time in the supposedly pure-bred Siamese litters born to a breeder in the United States. Although such kittens were discarded from breeding programmes at first, for not conforming to the usual Siamese standards, eventually it was decided to develop them as a separate variety. In the 1940s, two breeders in California and New York started serious work on the breed and today the Balinese is accepted in its own right.

An ideal Balinese has a fine-boned, svelte body with long elegant legs. The head is wedge-shaped with straight planes and the Oriental eyes are blue. The coat is long and silky, and as it has no woolly undercoat, it lies flat along the lines of the body. The ears may be tufted and the hair around the neck tends to be longer, giving the effect of an Elizabethan ruff. The hairs on the tail are long, making it plume-like.

As on the Siamese, the colour is restricted to the animal's points, namely the muzzle, ears, tail, legs and paws. Balinese are being bred in all the recognized points colours. The most popular in Britain, at present, is the Blue-point, and some charming Seal-, Chocolate- and Lilac-pointed cats have been seen. Work is progressing on the Tabby-point, and those with the red genetic factor – the Red-point, the Cream-point and their female counterparts, the Tortie-point. All these varieties are still classified as experimental in Britain.

Seal-point Siamese kitten Thairano Madison, son of Shandean Glastron

Lilac-point Balinese male – Cheldene Big Mac

In order to strengthen certain characteristics and to introduce new colours, as well as to prevent too much inbreeding, outcrosses have been made with normal Siamese cats.

The resulting offspring have short, rather plushy, Siamese-patterned coats and are known as Balinese Variants. All such Variants carry the long-coated genetic factor and are mated back to Balinese and in turn, produce true Balinese kittens.

Colourpoint

The Himalayan factor has been incorporated into several cat breeds besides the Siamese, and at present, a small group is developing British cats with the distinctive pattern under the temporary name of British Colourpoint, or Colourpoint Short-hair. The Colourpoint Longhair, however, is a truly Persian-type cat, with the addition of Siamese patterning, and must never be confused with the Balinese, which is a truly Siamese cat for both

colour and conformation, but with the addition of a long, silky coat.

Geneticists in the 1920s and 1930s deliberately crossed Siamese with other breeds to gain knowledge of the rules of inheritance in the cat. They soon realized the Himalayan factor is recessive to full coat colour, and that long hair in the cat is recessive to short hair.

Two workers at Harvard University's medical school carried out extensive trials, and were as concerned with conformation and structural characteristics as with coat colour. A five-year breeding programme with their carefully selected and well cared-for cats produced Debutante in 1935. She was a long-coated kitten with Siamese colouring, and was the first recorded Himalayan. Surprisingly, little interest was shown in the new variety at that time, and nothing much was done in the United States until 1950, when Marguerite Goforth, living up to her name, embarked upon a series of breeding experiments destined to produce perfect Himalayans.

America's leading registering body, Cat Fanciers' Association, granted recognition to the breed in 1957, and other associations soon followed suit.

Known as Himalayans in the U.S.A., and as Colourpoints in Britain, this charming pair are Princess Maritza, Blue-point female and Champion Frallon Creampoint Apache

Cream Colourpoint male

Mrs Goforth's reward came when her La Chiquita became the first Himalayan champion.

In Britain, the Experimental Breeders' Club was formed in 1935 and some Himalayan-patterned cats were produced, following plans formulated by Virginia Cobb in California. One of the first filial generation of pointed Longhairs was named Kala Dawn, and fortunately, she was kept safe and entire while war raged and cat breeding virtually ceased. When peace returned to Britain's shores, Kala Dawn was mated and produced a fine son, named Kala Sabu, and he was acquired by Brian Stirling-Webb, a pioneer in the perfection of new feline varieties. Mr Stirling-Webb collaborated with Dr Manton (Mrs Harding) and eventually the beautiful cats Briarry (a name derived from the breeders'

names) and Mingchiu were presented to the general public on exhibition.

The Governing Council of the Cat Fancy officially recognized the breed as the Colourpoint Longhair in 1955 and in 1958 the Best in Show award for a longhaired kitten at the Kensington Kitten and Neuter Cat Club show was taken by a Colourpoint.

Today, the Colourpoint has been bred in most of the colour varieties seen in the Siamese, including points colour of seal, blue, chocolate, lilac, red, cream, tabby and various shades of tortoiseshell.

The breed is typically Persian in type, with a cobby body, stocky legs and a short, full tail. The head is broad and round, with plenty of width between small, tufted ears, the cheeks are full and the nose short with a distinct 'stop' at the bridge. The large round eyes are a bright, clear blue. The coat has a distinctive soft texture and is long and thick with a full frill framing the face. The body colour is pale and the points stand out in clear contrast.

Just as the Colourpoint combines the colour and type of Siamese and Persian cats, so it appears to combine their temperaments, and the general characteristics of their behaviour. The female may precociously call as early as eight months, rather like a Siamese, but her litter will probably consist of only two or three kittens like that of a Persian. The kitten is forward, and playful, quick to learn, agile and alert, but less demanding and certainly less talkative than its Siamese cousins.

Birman

A Himalayan breed that has been in existence for many years is the Birman, or Sacred Cat in Burma, which differs from most other 'pointed' cats in having four striking white feet. The Birman is a longhaired cat of moderate longhaired type, being lighter in bone structure and without the extreme features of the main Persian varieties.

Believed to have originated in Burma before being developed as a breed in France, the first Birman cats were brought into Britain in 1965 and were accepted for championship status by the Governing Council of the Cat Fancy the following year.

A majestic cat, the Birman has its own unique standard, unlike the majority of the longhaired breeds which are of the same general conformation. The show Birman has a broad, round head with a sloping forehead and a flattened brow. Its ears are medium in size and wide at the base. It has full cheeks and a Roman nose and the blue eyes are rounded, but with a slightly Oriental look. The body is stocky but long, and is supported on fairly heavy limbs, while the tail is of medium length, nicely balancing the body.

This breed's unusual colour distribution is the subject of a fascinating legend from Burma. It is said that before the time of the Great Buddha, the Khmer people built beautiful temples in honour of their gods. One of these was the Temple of Lao-Tsun, built to the glory of the goddess Tsun-Kyan-Kse. It was here that the ancient high priest Mun-Ha officiated, constantly accompanied by his favourite among 100 pure white temple cats, a faithful feline named Sinh. One night the temple was raided by bandits and in trying to protect the statue of the goddess, Mun-Ha suffered a fatal heart attack and sank dying to the ground before the golden figure of Tsun-Kyan-Kse. At once Sinh jumped onto the head of his master hissing and growling in fury at the attackers and the junior priests rallied to repel the invaders.

Sinh relaxed and gazed into the sapphire blue eyes of the goddess and as he did so the perfect soul of Mun-Ha passed into the cat's body. A 'miracle' occurred as the transfer took place for the white coat of the cat became pale gold, like the reflected light from the statue; his yellow eyes turned to a matching blue and his face, legs and tail turned to the brown of the earth at the statue's feet.

Only his four feet, still protectively clasping his dead master's snowy hair, remained white. As peace reigned again in the temple, the priests discovered that all the other temple cats had undergone the same transformation as Sinh and from that day were considered sacred, as possessing the souls of departed holy men. On the seventh day after the raid, Sinh, who had refused all food and drink and sat looking steadfastly at Tsun-Kyan-Kse, passed peacefully away and to paradise, carrying with him the perfect soul of Mun-Ha.

Today's Birman cats still have the habit of staring long and hard into one's eyes and it can be a disconcerting experience.

The Ragdoll

The Ragdoll is a controversial breed, developed in the United States of America from descendants of a White Longhair queen. It is of semi-longhaired body type, rather like the Birman but larger and heavier, with a flowing silky coat. Three main colour varieties are accepted by some of the American Associations which have recognized the breed. First there is the Colorpoint Ragdoll, which is of standard Himalayan pattern – light body and darker points, allowed in the usual Himalayan/Siamese breeds. Secondly there is the Mitted Ragdoll, which is just like the Colorpoint, but with the addition of four stark white paws. Thirdly there is the Bicolor Ragdoll which is basically Himalayan – pointed

The Ragdoll, a pretty, if controversial, breed

caused great interest in the press, but the Governing Council of the Cat Fancy has not registered its approval of the scheme.

Other Himalayan Breeds

In America, even the Manx has been subjected to the influence of the Himalayan factor, thus producing the Si-Manx! British breeders are currently working on a well planned programme to develop the British Shorthair with points of all colours. Rex cats have been bred with the Himalayan gene and in the United States of America such cats are recognized in their own division as Si-Rex. In Britain, such cats must conform to either the Devon or Cornish Rex standards for type. Like all Himalayan cats, Si-Rex have clear points colouring and blue eyes.

BREEDS DEVELOPED IN AMERICA

The Burmese

Although many of our modern feline breeds were developed in Britain before being introduced to the United States, the Burmese is one of the few that made the trans-Atlantic trip in reverse.

It is a breed with an interesting genetic make-up, closely related to the Siamese and sharing the same genetic group known as the albino series. When a pure Burmese is mated with a pure Siamese, a beautiful intermediate variety, known as the Tonkinese in America, is produced. When Tonkinese cats are mated together, the offspring may be either true Burmese, true Siamese or Tonkinese.

In 1930, a retired ship's doctor, Joseph C. Thompson, made a trip to Burma. The San Francisco pedigree cat breeder was interested in all aspects of the East and when he was given an unusual little cat, he decided to take her home with him. The cat, called Wong Mau, was a Tonkinese and resembled a dark Siamese, brownish in colour with darker fur on her points. When she had settled down after the journey, she was mated to a Siamese, and her litter consisted of some Siamese kittens, and some just like herself.

Eventually some of her darker offspring were

with blue eyes, but has a considerable amount of white, usually on the legs, the chest, around the neck and extending up the throat, and over the muzzle to form a blaze, bisecting the face.

Breeders of Ragdoll cats allege that their pets are oblivious to pain, are totally fearless and will never fight. The name was given because the cats love to lie cradled in their owner's arms, and remain limp and floppy at all times, exactly like a toy ragdoll.

Recently, a number of Ragdolls were imported into Britain so that the breed may be developed, and

Blue Burmese – Champion Sidarka Delta Dawn

mated together and some of their kittens developed into the dark, glossy cats now called Brown or Sable Burmese. It was with the help of geneticist friends that Dr Thompson eventually deduced that Wong Mau was, in fact, a Burmese 'hybrid' or Tonkinese, and experiments confirmed that her darker coated descendants bred true. Thus the Burmese breed was born, and officially recognized in the United States in 1936.

Much confusion existed over the genetic implications of the Burmese/Siamese cross-matings and the expected variation in resulting litters. Eventually a famous paper entitled 'Genetics of the Burmese Cat', was published in 1943. It was compiled by Virginia Cobb, Madeleine Dmytryk and Clyde Keeler, working in close conjunction with Dr Thompson, who died just before its publication. The paper proved invaluable in the development of the breed and showed fanciers just where they were making their most serious mis-

takes. Cat Fanciers' Association, America's foremost registration body, concerned about bad breeding practices, withdrew recognition from the Burmese in 1947, but the efforts of the Burmese Cat Society of America were such that the breed status was restored in 1953. Other registering bodies had accepted the breed by this time and the brown cats gained in popularity.

During the breed's suspension in America, some Burmese were transported to England and soon established a paw-hold here. Mr and Mrs France of Derby introduced the Burmese to Britain, importing a male and two females in 1949 and another male in 1953. They were obviously a couple with tremendous foresight, but even they could not have had any notion of the impact the breed would have over the ensuing 30 years.

All the original Burmese cats were of a rich brown colour, although genetically black, but before long blue kittens began to appear in Burmese

Lilac Burmese queen – Champion Lilac Lana

litters when two cats carrying the elusive dilution gene were mated together. These kittens were considered to be particularly attractive and were developed as a separate variety, to be officially recognized by G.C.C.F. in 1960.

Possibly due to early Siamese cross-matings, some of the Brown and Blue Burmese were found to carry the chocolate factor, and when these were, by chance, mated together, Chocolate Burmese kittens were born. Then, when cats carrying both blue and chocolate genes were crossed, the rare Lilac Burmese came into being. The red factor was deliberately introduced into Burmese by outcrossing to Red-pointed Siamese, after a chance mismating between a Blue Burmese and a Red Tabby Shorthair resulted in a delightful Tortoiseshell of Burmese type.

Today the Burmese spectrum ranges from Brown (Sable), Blue, Chocolate and Lilac, through Red, Cream and Tortoiseshell to Blue-Tortie, Chocolate-Tortie and Lilac-Tortie. Unlike the situation in other breeds, the colour range recognized by most American groups is narrower than that in Britain.

In America the Burmese has become a substantial cat with a round head, full face and rounded eyes. In Britain, however, the standards call for a cat of a modified foreign type. Whatever its colour, the Burmese in Britain must be of a medium size, hard and muscular, with slender legs and a straight tail of medium length. The head should be slightly rounded on top, with wide-set, medium-sized ears, which, in profile, tilt slightly forward. The cheek bones are wide, giving the head a shortened wedge shape. The nose has a distinct break in the profile and the chin is firm and determined. The large and lustrous eyes should show an Oriental slant and for perfection should be golden-yellow, although other yellow shades are allowed. The coat is short and close-lying with a healthy gleam, and a Burmese in top condition feels surprisingly heavy when picked up.

The Bombay

The Bombay is a man-made breed of cat, developed in the USA and, as yet, not seen on the British show bench. The original cross was made between two Grand Champion cats, a Black American Shorthair male named Shawnee Anthracite, and a Sable Burmese (known as Brown Burmese in Britain) called Hill House Daniella. As neither cat carried the dilute factor, all the kittens were, as expected, jet black. They combined the type and hardy constitution of the Shorthair with the fine sleek coat and distinctive head of the American-style Burmese and provided a good basis for establishing the new breed.

Further breeding lines were produced and as genetically black stock was used throughout, no dilute or patterned kittens were ever produced. The desired traits were discussed and selected and before long the typical Bombay came into being and the planned characteristics were fixed in the variety.

On 1 May 1976, the breed was granted official championship status by the Cat Fanciers' Association. The C.F.A. standards are exacting and are aimed at the production of cats of consistently good type, conforming to the standard of points desired by fanciers.

The standard for the Bombay requires a cat of medium size and muscular conformation, neither too compact nor too rangy, and the males may be generally larger than the females. The legs must be in proportion to the body and the medium-length tail should be straight and free from any abnormality in its bone structure. The Bombay's head should be fairly large and pleasantly rounded without any flat planes, whether viewed from the front or in profile, but there should be a distinct and clearly visible nose-break and a firm chin.

The medium-sized ears have wide bases and rounded tips. They should tilt slightly forward and be set wide apart with a rounded top line to the head between them. The eyes are also wide-set with rounded apertures. Large and lustrous, the eye colour may be of any shade from gold to deep colour, although preference is given to the darker, more intense tones. Any sign of green in the eye colour is deemed a serious fault.

Short, fine and satin-like, the coat of the Bombay must be close-lying and have a reflective sheen which is more intensely black than that of any other black feline breed. Good feeding and the minimum of grooming keep the coat gleaming and coal-black, and the adult Bombay must have a coat totally sound to the roots. Kittens of the breed often have dull and unsound coats, but these improve gradually as the kitten-coat sheds, and the true coat develops with maturity.

Although the Bombay looks like a black American-style Burmese, the early pioneers of the variety thought it closely resembled a miniature version of the Indian black panther, and after much deliberation, chose the Bombay as its breed name.

Maine Coon

The Maine Coon is a breed of cat native to the state of Maine in North America, where it has been known as a pure variety for well over a hundred years. Although records of its past are a little sketchy, the first written record of the breed tells of a black-and-white longhaired cat owned by E.R. Pierce of Maine, which had the grandiose name of Captain Jenks of the Horse Marines.

The Coon was obviously very popular in its native state in the late nineteenth century, and it not only excelled as a rodent controller on the farms, but also made a perfect show cat. A tabby Coon was Best in Show at the Madison Square Gardens Cat Show of 1895, while a neutered Coon bred by E.R. Pierce won top awards. This neuter was called Cosie, and was a large brown tabby. Mr Pierce bred many winning Maine Coons which figured prominently in the show lists in Boston during 1897, 1898

A striking Bombay female at the Gemenee Cattery in San Diego

and 1899, but with the importation of other breeds from Europe, the Coon then suffered a decline.

Frances Simpson's *Book of the Cat*, 1903, devoted a whole chapter to the Maine Coon and related some of the tales told to explain the origins of the breed. One romantic story suggested that the prized longhaired cats of Queen Marie Antoinette were smuggled on board a ship bound for America, so that they could be conveyed to safety until the Queen could escape the French Revolution and rejoin them in safe exile.

A biologically impossible theory of the breed's origins was upheld for many years. This insisted that the cat was the result of matings between domestic farm cats and racoons, which produced the large size, dark tabby coats and bushy tails of the typical Maine cats.

The most likely explanation of the breed's origin is that attractive and unusual Angora cats were traded by New England seamen, and eventually, in typical feline manner, spread their longhaired genes among the native cat population. Thus the long-haired trait was combined with coats of all colours and patterns, although the dark tabby, being dominant, was the most often seen for many years.

From 1904 onwards, the breed's decline was dramatic, although a few devotees prevented its total extinction. Eventually, in 1953, the Central Maine Coon Cat Club was established and held specialist shows for the breed. Pedigrees were researched and careful breeding programmes were followed. The charm of the Coon captivated more and more fanciers, and in 1967, a standard was drawn up and accepted by some of the American registering bodies. In 1976, a new organization called the International Society for the Preservation of the Maine Coon was formed, and the Cat Fanciers' Association finally accepted the Maine Coon.

The Maine Coon cat has a unique and endearing personality, loving its owners, and liking to be near them at all times. It has a fondness for sleeping in high, strange places, on top of doors and cupboards, or draped along narrow shelves. It may eat its food with an elegantly hooked paw, and delight in scooping up water from its bowl in the same fashion.

Breed notes call for the Coon to have a head of medium width and length and showing a squareness to the muzzle. The high cheekbones emphasize large, wide-set eyes, which are gold, copper or green in all but white Coons, which have blue or odd-coloured eyes. The ears are large, and set high on the head, with beautiful tufts of silky hair. The nose is medium in length and the chin firm, lining up with the nose and upper lips. The whole head appears strong, lacking all extremes of type and functionally feline in all respects.

The Maine Coon is quite a large cat, but without any coarseness of bone. The body is long and well-proportioned with plenty of muscle, a broad chest and a good neck. The legs are sturdy and the paws round and well-tufted. The coat is full and quite heavy, shorter on the shoulders, but long and silky on the sides and flanks, and with a full ruff framing the face. The long tail provides the finishing touch to the Maine Coon, being covered with long, full and flowing hair, and is used constantly in communication.

On the show bench, the Coon is penalized for having a short or even coat, delicate bone structure, a squint or a kinked tail. It may however, be of any coat colour or pattern, so there is a Coon cat to suit anyone's taste.

It is interesting to compare the breed with the Norwegian Forest Cat, a uniquely Scandinavian feline. This cat is almost identical to the Maine Coon, except that its hind legs are proportionately longer and it has a slightly thicker coat during the colder months of the year.

TURKISH CATS

Two breeds of cat are accepted in Turkey, and one of these, the greatly prized Van Cat, has become established as a distinctive longhaired breed in Britain, where it is recognized as the Turkish Cat. In America, Turkish cats are known officially as Van Cats.

It was during her travels in the Lake Van area in south-eastern Turkey that Miss Laura Lushington first met cats of this unusual and enchanting type. In

Odd-eyed White Turkish Angora kitten – Kartopu

1955, she was lucky enough to be presented with a pair of kittens, which she eventually brought back with her to England.

In their native home, the Van Cats are loved for their intelligence and exceptional strength of character. They are affectionate and active, and possess an unusual colouring, being mainly white with dark auburn markings. Perhaps their most outstanding characteristic is their apparent love for water, for Van Cats not only like to dabble their paws into pools but have also been known to jump into lakes and streams. This gained them the nickname of 'swimming cats'.

Miss Lushington made several trips to Turkey, each time bringing home carefully selected breeding stock of Van Cats, all of which had to undergo the statutory six-month period of quarantine on arrival in England. Then she started on the long road towards official recognition by the G.C.C.F., which in those days entailed scientific breeding and registration of four generations within the same variety. It says much for her determination and fortitude that in 1969 full pedigree status was achieved – Turkish cats had arrived, were eligible for Champion Challenge Certificates and the British Cat Fancy had a new, natural breed.

Today's Turkish must not be confused with the Turkish Angora, an ancient breed in the process of being re-established in Britain as the Angora cat. The Turkish retains all the characteristics of its Van ancestors, and for show purposes must meet an exacting standard. The coat is of long, soft and silky fur with no woolly undercoat. It must be chalk white in colour and free from any tinge or trace of yellow shading. There should be dark auburn markings on the face, leaving a white blaze and white ears. The tip of the nose, the paw pads and the inside of the large, upright ears should be delicate shell pink.

The Turkish may have small, irregularly placed auburn markings on the body without penalty, and the full, brush-like tail is completely auburn, strongly marked with tabby rings. The cat is sturdy, and of medium to large size with a long muscular body and strong legs. The rounded paws have tufts of fur between the toes. The head has strong lines and is a short wedge in shape with a fairly long nose and good width between the distinctive light amber-coloured eyes which are large and round, with pink rims.

Angora cats are being re-bred in Britain, and are therefore in the 'experimental' category as far as the G.C.C.F. is concerned, having been granted preliminary recognition in 1977. In the United States, however, the Turkish Angora is a well-established breed and is accepted in a wide range of coat colours including White, Black, Blue, Black and Blue Smoke, Red, Brown, Blue and Silver Tabby in Classic or Mackerel patterns, Calico and Bi-color.

The Cat Fanciers' Association standards call for a cat of medium size with a long lithe body and lengthy limbs with small tufted paws. The medium wedge-shaped head should be wide at the top with a definite taper towards the chin, a medium, gently sloping nose and long pointed and tufted ears set high on the head. The large eyes should be round or almond-shaped and should tilt slightly at the corners. They should be amber to complement most coat colours, but White Angora cats may have blue, amber or odd eyes, and the Silver Tabbies may have eyes of green or hazel.

Silky, and with a tendency to wave, the coat of the Angora is medium in length over the body, wavier on the stomach and long around the neck, giving a full ruff. The long and tapering tail is also furnished with extra long hair, and is often carried horizontally over the body, the tip almost touching the cat's head.

MANX CATS

The true Manx cat is unique in being totally tailless, and is an ancient breed whose origins are shrouded in mystery and legend. One story relates the wrecking of the Spanish Armada off the English coast in 1588 and suggests that tailless cats swam to safety on the Isle of Man. Another legend insists that warriors living on the island favoured cats' tails as plumes for their helmets and to prevent the later mutilation taking place, mother cats would bite the tails off their new-born kittens. The Celts believed that treading on a cat's tail caused a viper to appear, using its venomous bite, and so cats developed without such evil-provoking tails.

Champion Sunacres Rosey Dawn, a red-and-white Manx with winning ways

Even Noah has been blamed for the Manx cat's rump, for it is said that Noah was just about to close the door of the great ark, despite the fact that the pair of cats had failed to arrive. Just then, the cats strolled leisurely aboard, and the door swung shut, trapping their tails.

It has also been said that the Manx resulted from the crossing of a cat with a rabbit, but this is biologically impossible. In fact there are short- and stubby-tailed cats in the Far East and it is more than likely that such cats were transported to the Isle of Man by the Phoenician sea traders, where the confines of the island environment ensured that sufficiently close breeding would occur to establish the tailless cat as a fixed strain. The original

mutation could even have occurred on the island itself, but such cats were first recorded there in 1820.

A Manx Cat Club was formed in 1901, and it has been said that King Edward VII kept several of the breed as pets when he was Prince of Wales. During the 1930s the breed declined in popularity in Britain, although it gained favour in the United States, and even today, the Manx is only seen in small numbers on the show benches of the world. The main reason for the scarcity of the Manx is that its tailless condition is due to a semi-lethal genetic factor which may affect other parts of the body as well as limiting the end of the vertebral chain. Sometimes fusion occurs in other parts of the spine, or the distressing condition known as spina bifida may manifest itself.

The fit and healthy Manx admired at shows is a heterozygote, which means that it has one gene for taillessness and one gene for normal tail, but because the former is dominant to the latter, the cat has no tail.

When a Manx to Manx mating occurs, one in four of the conceived kittens is likely to be homozygous for taillessness, and dies in utero or shortly after birth. The following simple diagram shows how this comes about.

The 'normal' heterozygous Manx has the tailless gene M and the normal tail gene m, therefore its kittens will receive either an M gene or an m gene. When two such cats are mated:

	M	m
M	MM (usually dies in utero)	Mm (Manx)
m	Mm (Manx)	mm (tailed kitten)

Manx: Mm

Note: Mating two Manx together usually results in the birth of a 2:1 ratio of Manx to tailed offspring. The homozygote Manx probably dies in utero, due to a lethal genetic factor.

The kitten with the genotype MM is homozygous for taillessness and dies, the two kittens with genotype Mm are 'normal' Manx cats, while the kitten with genotype mm has no gene for taillessness and therefore has a tail.

As the tailless factor often affects the whole of the vertebral column, Manx cats have other distinguishing physiological features. The forelegs are slightly shorter than normal, and the hind legs are particularly long, the back is quite short and the rump high and rounded.

There are degrees of taillessness and the true Manx, known as a Rumpy, has a small dimple or indentation where the tail should join the body. The Rumpy-riser, as approved by the Cat Fanciers' Association in the United States has a small rise of bone at the end of the spine; this is deemed a fault in G.C.C.F. show specifications.

Manx cats with short tail stumps are called Stumpies in Britain and Stubbies in America and are often favoured by breeders along with the Longie, or Long-tailed Manx. These cats are usually preferred over standard short-haired out-crosses because the Manx show cat has a definite and specified head and body type.

A top show quality Manx should have a head similar to that of the British Shorthair, round and with prominent cheeks, but with a straight and longish nose. The ears are longer, and set higher on the head. The body is solid and compact – the whole cat looking as though it would fit comfortably into a square box.

The Manx may be of almost any normal recognized feline coat colour or pattern, and its large lustrous eyes should be of a shade which complements the coat. The breed is shorthaired, but in the 1960s some long-coated kittens were born in America, and gave rise to a separate tailless breed known as the Cymric.

In 1960, the Manx Parliament set up a state-owned cattery on the island, which was taken over by the Douglas Corporation in 1964. The cattery was designed to preserve the breed in its pure state and is open to visitors to the island. The cats kept there are not of the extreme type as seen on the show bench but are happy, healthy animals, seemingly content to preserve their heritage.

THE SCOTTISH FOLD

In 1961, a young kitten was seen roaming freely in a Scottish farmyard, and was particularly noticeable because its ears were folded and bent forward over its forehead instead of being pricked in the normal

way. When she matured and had kittens of her own, William Ross, a kindly shepherd who cared for her, saw that two of her youngsters had the same strangely folded ears. He contacted geneticists and breeders and following their advice, decided to carry out a breeding programme aimed at the production of further folded-ear cats.

The Scottish Fold resembles a British Shorthair in its head and body shape, but is unique in having the small ears tightly folded and set in a cap-like fashion, giving a totally round appearance to the cranium.

The folded ear effect is caused by a simple dominant genetic factor, so only one parent needs to have such ears in order to produce folded-ear kittens. Allied with the factor is the tendency to produce thickened tails and limbs, and because of

Capricorn's Gilmour of Swady, a Scottish Fold recognized in America but not in the country of its birth

this, the Governing Council of the Cat Fancy in Britain decided to refuse registrations for Scottish Folds, in effect banning their breeding. Many robust Fold kittens with regular bone structure have been sent to the United States of America however, and by breeding the best Fold stock to top-class American Shorthairs, the breed is now firmly established and a popular pet.

The Cat Fanciers' Association granted full championship status to the Scottish Fold in 1978, recognizing it in all regular coat colours and patterns except chocolate, lilac and Himalayan-pattern.

REX CATS

The Rex Coat

The normal coat of the domestic cat is made up of three different types of hair. There are guard hairs, bristle or awn hairs, and wool or down hairs.

Guard hairs, along with awn hairs, are often collectively referred to as the topcoat or overhair, while the down hairs are called the undercoat or underfur. The topcoat acts as a protective covering to the softer underfur, which is, in effect, an insulative layer. When viewed under a magnifying lens, the guard hairs are seen to be straight, tapering evenly to finely pointed tips. The awn hairs are thinner than the guard hairs and swell noticeably near their tips before tapering to fine points. These hairs vary in appearance and may be graded into three types, some approaching guard hairs in form, while others are as thin as down hairs. The down hairs are the shortest and thinnest of all the types and appear to be of similar diameter along their length. Tiny undulations in their structure cause them to be slightly crimped. Most cats have all three hair types present in their coats in standard proportions, but Rex-coated cats are decidedly different.

The two British Rex breeds are the Cornish and the Devon, both with characteristically curled coats, but each caused by a separate mutant gene.

In the Cornish Rex the very few guard hairs which might be present are generally reduced to resemble awn hairs. Occasionally, the awn hairs are sparse, or missing altogether, and the cat then has an extremely short, fine, soft coat which feels remarkably silken to the touch. The coat is even shorter on the extremities, and feels like soft velvet pile.

The Cornish cat has deeply defined and rippling waves, arranged in well-ordered regular patterns, over its body. Most Cornish Rex are densely coated and are penalized on the show bench for showing any bare patches and straight or shaggy coats. Their whiskers and eyebrows often make long, graceful curves, adding to the breed's quizzical charm.

The coat of the Devon Rex is coarser than that of its Cornish cousin, feeling more like suede than silk. This effect is due to the fact that the guard hairs, while not being eliminated by the Devon gene, are modified in a unique manner. They are similar in structure to awn hairs but are uneven along their length and often have broken, split ends. Individual cats show noticeable variation in the length of the guard, awn and down hairs, and Devon Rex shed their coats dramatically during the moulting process. This has the effect of producing cats with well-furred heads, legs and tails, but rather bare bodies. In full coat, the Devon may have well-ordered waves along the back, though these are never as clearly defined as in the Cornish, and become disarranged during periods of normal activity. When an abundant coat is present, it tends to form ringlets on the abdomen rather than rows of waves.

Some Devons have dense curled hair at the outer edges of their ears. This characteristic is called 'earmuffs' and while not specifically required by the standard of points, is not penalized by judges. Many fanciers selectively breed for lynx-like tufts at the tips of the large ears, considering that these enhance the breed's pixie-faced look. The whiskers of the Devon Rex are extremely brittle, breaking easily and leaving bent stubs. If they do achieve any length, they are usually kinked or twisted.

Cornish Rex

For some years local country folk had heard tales of curly-coated cats living wild on Bodmin Moor in Cornwall, but it was not until the summer of 1950 that such a cat was born to a domesticated queen, unknowingly creating a milestone in feline history.

Mrs Ennismore lived in an old farmhouse on the moor and her pretty little Tortoiseshell-and-White shorthaired queen Serena gave birth to a litter of five kittens. Four of the kittens looked normal, but one, a cream male, appeared to have a waved coat. At first Mrs Ennismore thought this phenomenon was due to the wetting effect of the birth secretions, but as the kitten dried out the curls remained. The veterinary surgeon suggested that Mrs Ennismore should write to the geneticists Jude and Searle. The kitten's hair structure was examined under a microscope and identified as being similar to that of the Rex rabbit. It was on the geneticists' advice that the curly kitten was called a Rex cat.

Oriental Silver Spotted Tabby

Silver Tabby Longhair

Male and female Somali

Shaded Silver American Shorthair

Lilac-Cream, Chocolate-Cream and Blue-Cream Burmese

Chocolate-point Siamese kitten

Lilac-point Siamese

Best in Show pedigree kitten

Seal-point Birman

Newborn litter

When he matured, Kallibunker, as the Rex was called, was mated to his mother, and in the resultant litter there were two rex-coated kittens and a flat-coated one. The mating was repeated later, and further rex-coated offspring were produced. Unfortunately Kallibunker died quite young, but one of his sons, Poldhu, carried on the line and was the sire of Lamorna Cove, a female exported to the United States where she in turn founded a strong Rex family.

Although Poldhu is said to have been a Blue-Cream male, the evenness of his markings and the colour record of his offspring strongly indicate he was most probably a Blue-Tabby.

As fertile Blue-Cream males are extremely rare, it was decided that a tissue sample be taken from Poldhu, presumably for chromosomal examination,

and in having this operation performed, the cat was, unfortunately, rendered sterile. To make matters even worse, the important tissue sample was mislaid. Luckily another of Kallibunker's sons was still entire, owned by Mrs Rickard, Mrs Ennismore's veterinary surgeon and friend.

It was decided that this cream-and-white cat should be loaned to Brian Stirling-Webb who was conducting planned breeding experiments to instill extra stamina and the propensity for longevity into the breed, as well as attempting to standardize type. Registered as Sham Pain Chas, this cat was mated to three British Shorthair queens and one Burmese, and an enthusiastic group of breeders thus started along the arduous road to producing a viable, recognizable feline variety.

The Rex gene responsible for causing the curly-

Amaska Highlights, a red Cornish Rex, was born in Britain but now lives in Cincinnati

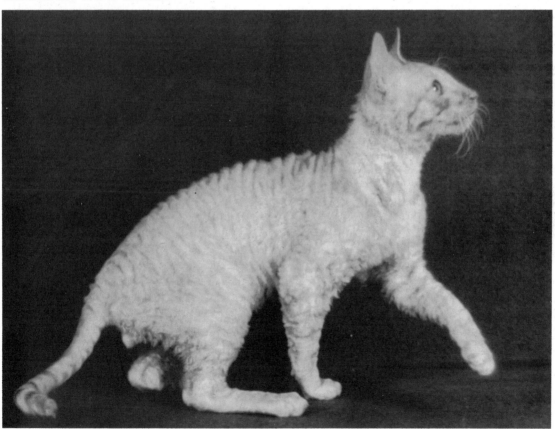

coated effect of the Cornish Rex cat is recessive to the gene for a normal flat coat, which means that both parents have to carry the gene for it to be apparent in their kittens. Because of this, all the kittens born to plain-coated shorthaired queens had flat coats, but as their sire was Rex, every one carried the special Rex gene, and when they matured and were mated to one another, or back to their father, some of their offspring carried the elusive curled coat.

Records show that the original Rex cats of Bodmin Moor varied considerably in type, some being dainty and fine boned and others of much stockier build. Breeders worked hard to produce Cornish Rex of standard conformation. Some lines were bred from the original stock mated with Havana and other Shorthairs, and some were developed from descendants of Lamorna Cove, reintroduced to Britain from Canada.

Today the Cornish Rex is a popular pedigree cat which always causes ripples of interest amongst visitors to cat shows, echoing the ripples of tight marcelled waves along its beautiful coat.

The breed standard calls for a hard and muscular cat of slender build with long straight legs and small oval paws. The tail should be long, fine and tapered without any kinks. The distinctive head is wedge-shaped and one-third longer than it is wide, with a straight profile, a firm chin and large, wide-based ears set rather high on the head. The oval eyes should be of medium size and their colour should be in keeping with that of the coat. As one would expect, the coat is an important show feature and carries high marks. It must be short and plushy in texture without any guard hairs. It should wave, curl or ripple over the body, the legs and along the length of the tail, and even the whiskers and eyebrows tend to curl.

This breed may be of any accepted feline colour, and if white markings are present, these must be symmetrical except in the case of the Tortoiseshell-and-White, when they may be randomly placed.

Devon Rex

Ten years after the discovery of the first Cornish Rex kitten at a farm on Bodmin Moor, the breed received a great deal of publicity with photographs and articles in the national press. The *Daily Mirror* printed a charming picture of Du-Bu Lambtex, an enchanting blue-and-white kitten, captioned with details of her unusual curly coat. This article was read with avid interest by Beryl Cox, who lived near an old tin mine in Buckfastleigh, Devon. A colony of feral cats inhabited the old mine workings and were impossible to catch or tame. One dominant male was particularly elusive, but when he was glimpsed, it was apparent that he had an unusual coat and a tail which appeared to be covered in ringlets.

One day Miss Cox discovered a stray cat in the hedgerow at the bottom of her garden. The little female had just given birth to a litter of kittens, and these were quickly gathered up and taken, with their weakened mother, into the house. One of these kittens had a strangely waved coat and was kept as a pet by Miss Cox after his litter-mates had gone to new homes. Miss Cox called the kitten Kirlee, and after telephoning the newspaper for information about Rex cats, was able to contact Mrs Watts, the breeder of Lambtex, to talk about her unusual pet.

Kirlee was unusual in appearance. As well as having a strange coat, he was a smokey mole-grey colour and the thicker areas of fur on his extremities gave him an almost 'pointed' appearance.

Extremely affectionate and highly intelligent, Kirlee spent a happy kittenhood with Miss Cox. He followed her everywhere – just like a little dog – and oozed personality. His favourite trick was to walk a thick taut rope, wagging his tail from side to side to aid his balancing act.

Eventually, in 1961, when he was nine months old, Miss Cox allowed Kirlee to join the Rex breeding programme and Brian Stirling-Webb arranged matings between this young cat and some of the existing Rex-coated, and Rex-carrying queens. To everyone's surprise, every kitten resulting from these matings proved to have a flat coat, and it eventually dawned on the bemused breeders that Kirlee's curls must have been due to the effect of a totally different Rex gene.

After expert investigation it was decided to christen the genes Gene 1 for that which produced the Cornish curls, and Gene 2 for that which was responsible for the Devon cats. It was rather unfortunate that by the time the Rex genetics had been sorted out that there was quite a population of cats carrying both Gene 1 and Gene 2 in their makeup, and a time-consuming programme of test-matings had to be carried out in order to determine just which genes were carried by each cat so that the two breeds could be developed quite separately, with the Gene 1 animals being called Cornish Rex, and the Gene 2, Devon Rex.

The Devon Rex is a cat of medium build with a hard, muscular body and a characteristically broad chest. Its legs are long and slim, ending in small oval paws, and the tail is long, fine and tapered. The head is shortish and wedge-shaped with full cheeks, and in profile there should be a strongly defined stop at the bridge of the nose. The Devon's ears are large and wide-based, set fairly low on the head and often have a profusion of curled fur giving the appearance of ear-muffs. The eyes should be oval in shape but large and wide apart. They may be of any colour, so long as they are in keeping with that of the coat.

Devon Rex show cats are faulted if they show

Garchell Dopey Dreamer, a delightful Devon Rex whose looks belie his name

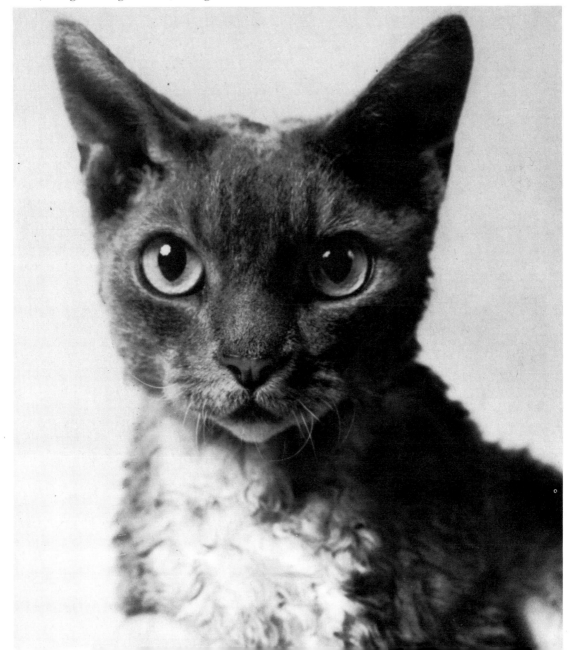

straight or shaggy coats, or if the tail is short, bare, bushy or kinked. Bare patches on the body are a serious fault, but many Devon Rex cats which appear bare on the underparts are seen on closer examination to be covered in fine down.

The typical Devon coat, being devoid of any protective guard hairs, soon soaks up water, so this breed should be kept out of the rain and covered runs are advisable in the cattery. Young Devon kittens are appealing, with their huge bat-like ears and heavily wrinkled brows.

There was a move some years ago to call the breed the Butterfly Rex because, viewed from behind, the ears did resemble the spread wings of a butterfly. The typical broadness of the chest some-times causes the forelegs to appear rather bowed and this, combined with a natural quizzical expres-sion, gives a pixie-like look.

Other Rex Cats

Rex cats have appeared in other countries besides Britain. The German Rex and the Oregon Rex were found to have coats similar to that of the Cornish cat. The German variety, however, was found to have thicker awn hairs, while the Oregon cat's awn hairs were thicker still. In both types the awn hairs bent and curved so much they did not project above the down fur as in normal coats. In recent years yet another Rex mutated, this time in Holland and was found to be quite different – the gene involved acting as a dominant factor to normal coat, while all other Rex types had proved to be recessive.

Dutch Rex kittens are born with tightly curled coats, but the effect gradually lessens with age, the waviness being replaced by a fairly coarse, bristly pelage. The guard and awn hairs are much thinner and shorter than in a normal feline coat, and are so bent and waved that they protrude in random directions above the general level of the underfur.

Breeding

THE STUD MALE

While a good quality brood queen plays a vital part in any pedigree cat breeding programme, the stud male has an even more important role. A fertile queen will produce anything from 30 to 100 kittens in her lifetime, while the male may sire hundreds of kittens to many different queens, so a male cat kept for stud purposes must be an exceptional specimen of his breed. He must be selected for his good looks which should conform closely to the standard of points laid down for his breed, and he must also have other qualities. He should be well-reared and have been free from any illnesses during kittenhood and as an adolescent; his parents should be sound and both from lines known for high fertility. Most important of all, the stud male must have a well-balanced, even temperament so that he will enjoy being handled by strangers. This ensures that he will show himself to advantage on the bench and endear him to the judges, as well as saving him from becoming upset when visiting queens show aggression.

Keeping a male cat at stud is strictly for the fancier with years of experience in showing and breeding who has gained the knowledge required for the intelligent keeping of an entire male cat. It is also important to understand both the psychology and the physiology of the cat in order to be successful at stud work. Entire males are generally very affectionate but any unfortunate cat kept in unsatisfactory or cramped surroundings will become morose, touchy or withdrawn. Some male cats are kept by those who consider the regular stud fees as a welcome extra income, while others, once famous champions and in full work with two or three queens each week in the breeding season, find themselves deposed in favour of new young champions. Instead of having such cats neutered, their owners keep them entire, shut away in solitary stud houses, awaiting the very occasional visiting queen. Such cats are usually well fed, warm, clean and tidy, but spend their days in bored and frustrated loneliness.

The stud male must be housed safely and comfortably, and each owner has his own idea of cattery construction and site, but many cats are kept too far from the house and soon become bored, when they may start to pace the run and wail. The best site for a stud's house is near plenty of regular activity so that he can see what is going on. The stud house can be of virtually any construction; sheds of all shapes and sizes, summer-houses, kennels, outhouses, stables, workshops, conservatories have all been successfully converted for stud work. The basic requirements call for a building that you can stand up in, and large enough to enable you to attend the male and his visiting queen even in the most inclement weather. It should be large enough for the male to exercise indoors through the long winter months when the outside run may be too cold, wet or draughty. The walls, floor and ceiling should be lined for warmth and painted with an impervious, easily cleaned material, or covered with laminated board. The floor must also be adequately covered and easy to clean. There should not be any crevices to harbour germs or parasites, and there should be a pen or cage to contain the visiting queen while she is getting to know the stud. Some stud houses have very large pens and runs for visiting queens but as a female is at stud for a very short time she requires only a small area in which to settle and feel more secure.

The stud, on the other hand, lives permanently in his quarters so he should have all the space that is available. Some form of heating is essential in the stud house and it is often better to spot-heat over the sleeping area with suspended infra-red lamps, than to attempt to heat the whole interior of the building. Stud males must be kept comfortably warm, but should never become used to very high living temperatures or their coats suffer and they may become susceptible to low-grade upper respiratory conditions.

It is important that the queen's pen is cosy and warm, for the visiting female is stressed by the change in environment and often refuses to eat or drink for 24 hours until she settles in. Extra warmth at this time is essential. The run should be as large as possible and high enough for the attendant to enter without stooping. Grass looks fine when freshly planted but it is difficult to cut within the run and virtually impossible to sterilize between visiting queens. Stud cats usually develop natural territorial marking habits which may include the faecal marking of the run as well as urine-spraying along the perimeters. If the run is concreted or paved it may be washed with a solution of bleach which keeps it fresh, clean and free from odours. Shelves and logs are excellent for exercise and sunning and the roof and walls must be securely wired in so that they are totally escape proof. All doors to both house and run must be close fitting and have secure fastenings.

Stud cats should be groomed and handled as much as possible and kept in peak condition by intelligent feeding and should be encouraged to exercise well, with toys and games between visits from queens. Vaccination programmes must be instituted and regular bloodtests taken to ensure freedom from Feline Leukaemia Virus. All possible parasites should be controlled. Strong disinfectants should not be used around the stud house, the floor should be mopped each day and if two or three fresh

toilet trays are provided there should be no 'tom-cat odour'. A row of picture hangers may be fixed around the stud house walls to hold three large bulldog clips which in turn hold several sheets of newspaper, hanging down the wall to touch the floor. These form a perfect spray barrier and can be changed when necessary; they also protect the wall from splatters around the toilet trays and save a great deal of work. The best way of cleaning a stud house is to wash the walls and floor thoroughly while the male is out in his run, then to wipe the surfaces with a solution of bleach, made up to the manufacturer's instructions, applied with a sponge mop. When this is thoroughly dry it is safe to allow the cat back into the building.

It is prudent to have a complete set of cleaning materials hanging in the stud house – broom, dustpan and brush, mop, kitchen towel roll and holder and a covered wastebin. This saves time and prevents any chance of cross-infection when several cats are housed in the cattery. When the run is swept, the bleach solution remaining in the bucket can be used up by brushing it over the mesh walls to cat height, which helps to deter the male from spraying the perimeters of his run. Most males spray copiously when they first take possession of new stud quarters, but usually settle down quickly and many males stop spraying after becoming well-established and secure within their territory.

As soon as a cat loses his popularity as a stud he should be neutered, and even the most virile of working cats is capable of turning into a perfect pet. Unfortunately, some cats show lapses in their manners after the operation and may spray indoors, so should only be allowed in rooms with washable furnishings and floors. Ex-stud neuters retain their male musculature and typically hard condition if fed intelligently and not allowed to get lazy, and male cats who have spent years uttering threats to each other across the cattery often become inseparable friends after neutering.

Keeping a stud cat can be very rewarding for it is often possible for an owner to build up a rapport with an entire male which lasts for his entire lifetime. It is unlikely to be financially rewarding however, for the costs of feeding, heating and housing generally outweigh the income derived from stud fees. If you have several queens of your own, there are distinct advantages in having your own stud cat for the queens conceive more easily when not subjected to travel and strange surroundings.

THE QUEEN

The entire adult female cat is called a queen and depending on her physical state and breed she may be in breeding condition for most of the year. Peaks of sexual activity are seen to occur in queens however, usually in the early spring and again in late summer. Queens which are kept indoors, living at fairly constant temperatures, and subjected to long hours of white light conditions, breed more frequently than those kept in very natural conditions.

In nature kittens born during the warmer months of the year obviously stand a better chance of survival. Spring and summer born kittens tend to be more robust and seem to have a natural resistance to most diseases, whereas winter born kittens may be frail. Adolescence varies in the female cat. Some precocious kittens may call as early as four months, while others come into first season as late as 15 months. The signs of oestrus are easy to recognize and consist of four quite distinct stages which occur in a cyclic pattern. The queen in oestrus is described as being 'in season', 'on heat' or 'calling'. The preparatory stage is known as *pro-oestrus* and is the period during which the reproductive organs undergo subtle changes in readiness for mating, fertilization and pregnancy. During this stage the queen becomes very affectionate and rather restless. If confined to the house she paces the floor and spends long periods gazing out of the window. She is seen to rub the scent glands along the side of her lips against furnishings and on the floor and may roll from time to time. Although she may show interest in a male cat at this stage she will not allow mating to take place. Pro-oestrus may last from one to five days.

The second stage is known as *oestrus* and this is the period during which the queen will allow mating. Oestrus lasts for about seven days, during which the cat becomes increasingly agitated, rolling vigorously on the floor and crying constantly. She may try to escape if confined and will even recklessly jump from upstairs windows. Her cries are deep and urgent and she may refuse to eat. If stroked firmly along her back, the queen should take up the typical mating condition, tail to one side, hindquarters raised, hindlegs extended and her chest and head held low. The queen may be very resentful of being lifted or handled. The frustrated queen may behave unsocially around the house, neglecting to use her toilet tray and spraying urine on the furnishings and walls.

The third stage in the cycle is *metoestrus* and occurs when the queen has not been mated and fertilized during oestrus. During this stage the reproductive system enters a relaxed state. The stage lasts about 24 hours during which the queen will still accept the advances of the male but may be reluctant to let actual mating take place.

The final stage is *anoestrus* and is the resting period. The queen is quite relaxed and totally disinterested in males. Anoestrus can last from ten days to several months and though the duration varies considerably with individual queens, each female appears to have a fairly consistent cycle of her own.

In the United States of America, queens are often mated during their first period of oestrus if this occurs after the cat has reached the age of ten months. In Britain, however, queens are generally allowed to go through the first period of oestrus and are carefully observed to note the calling pattern. They are then mated on their second period of heat.

The responsible owner will arrange for the queen to have a complete pre-mating checkup by the veterinary surgeon before taking her to stud. She may also require a blood test to ensure that she is free from Feline Leukaemia Virus. A sample of the queen's faeces should be handed over so that it can be checked for parasitic worms. Such a sample should be collected from the queen's toilet tray and placed in a clean glass jar or polythene bag. The sample is analysed under a microscope which allows accurate identification of any worm eggs present. During the physical examination the veterinarian checks the queen for other parasites such as fleas or earmites. The vaccination certificates should be checked too, for any boosters must be given well before mating takes place. The female is also examined to make sure that she is physically ready for pregnancy and birth.

Before mating, the queen should be given an adequate and well-balanced diet, including some raw meat to keep her teeth and jaws healthy. She should be regularly groomed and kept in fairly close confinement to ensure that she does not come into contact with any outside, and possibly infectious, cats.

MATING

It is usual to contact the stud male's owner as soon as the female comes into oestrus. Then arrangements may be made to transport the queen to the stud on the second or third day. After the visiting queen has had time to settle in her quarters within the stud's house, mating may take place. The settling in period may last for a few hours or two or three days. If the queen appears to have gone off call, she should be removed from the stud house at feeding times so that the male cat can rest and feed. He will be unlikely to accept food while the queen is in the pen. Happily, most queens settle in fairly quickly and within ten to twelve hours of their arrival croon to the male and take up the mating posture. When this occurs the queen can be released from the pen so that the stud can approach her. The stud's owner should stay, keeping perfectly quiet and still but not interfering, to watch proceedings. The male approaches the queen from the side and rear and grasps the loose skin at the scruff of her neck in his jaws. Placing one foreleg on either side of her shoulders he gently kicks at her flanks with his hindlegs, encouraging her to raise her quarters. This positioning procedure may go on for some time, but if the male is very experienced, penetration may occur quite quickly. As penetration takes place, the queen growls fiercely, and tries to escape from the male's grasp. Ejaculation is quite rapid, after which the male releases the queen and jumps away from her, as she turns to strike him. The queen rolls vigorously for a few moments then both cats sit and clean their genitals. Both cats will be ready to mate

again within a few moments and unless controlled, will continue to mate at intervals for several hours. Most stud owners allow three matings daily for two to three days before the queen is returned home.

THE PREGNANT QUEEN

The Gestation Period

The time from conception to parturition is known as the period of gestation, and in the cat this averages 65 days. Cat books generally state slightly different numbers of days for feline gestation, but experienced breeders agree that the 65th day is the most common for birth to take place and having counted nine weeks from the date of the first observed mating, start watching the queen carefully for the early signs of labour.

Kittens have survived which arrived only 59 days after conception, although this is rare, and many queens carry their litters longer than the average, giving birth on the 67th–69th day. It must be remembered that conception occurs *after* mating and may be delayed so it is possible that these supposedly overdue babies were, in fact, to exact term. If a queen is eating properly and seems well, there is no need to panic when the 65th day comes and goes with no sign of labour, and there is no need to have her examined by the veterinary surgeon until the 70th day unless she seems very lethargic, is hot to the touch, loses her appetite or shows any unpleasant vaginal discharge. Throughout gestation the queen should be treated quite normally, for pregnancy and birth are completely natural functions. She should not be coddled or treated with undue sentimentality for it has been found that over-humanized queens generally prove to be unsatisfactory mothers. A spoiled queen may refuse to feed or care for her kittens unless her owner stays constantly at hand. She may produce rather acid milk resulting in kittens which are consequently undersized and prone to gastric problems, or she may resent the interest in her litter and spend hours carrying the hapless mites in and out of the nest box, perhaps lacerating their necks as she carries them to and fro, which could in turn lead to serious infection and even death.

Weeks One to Three

It may be difficult to tell whether or not a queen is pregnant for the first two or three weeks after mating, so she should be treated quite normally and fed her usual favourite diet. The first hopeful sign that she may be safely in kitten generally occurs three weeks after conception, when a careful inspection of her nipples will show them to be distinctly pink in colour and very slightly enlarged. This change is particularly apparent in maiden queens whose nipples have until this time been pale and very tiny. In the older queen, especially in one that has had several litters, the nipples are perpetually enlarged, and may often look pink-toned towards evening whether she is pregnant or not. This phenomenon, termed 'pinking up' by most breeders, is caused by hormonal changes in the queen. When pinking-up is observed in conjunction with a relaxed attitude of the cat, and no signs of a further period of heat, it is fairly safe to assume that kittens may be expected in about six weeks.

Some queens suffer from morning sickness and are inclined to vomit white or pale yellow frothy liquid. Such cats must be watched carefully to ensure that they have no other symptoms which could indicate an infection contracted while at stud. The diet may be changed slightly, so that easily digestible foods are given for a few days, and the time of the evening meal may be brought forward considerably. If the queen seems subdued or runs a temperature, then veterinary advice should be sought.

Week Four

During the fourth week of pregnancy the queen's abdomen may appear to be very slightly swollen and a veterinary surgeon may be able to feel the tiny embryonic kittens by gentle palpation. At this time, the kittens are distinguishable as small distinct bumps within the uterus. The inexperienced should never attempt this test for themselves for careless probing or kneading could cause internal damage to the queen or induce an abortion. It is at this stage that the cat's appetite increases very slightly and she seems to appreciate the chance to chew on grasses or herbage. She seems to settle down and become more

calm and relaxed in her general demeanour. If she has been nauseous in previous weeks, this stops and her body functions settle down again.

Week Five

The queen's abdomen now becomes visibly swollen whether she is moving or in repose and in all queens, the nipples show a definite degree of enlargement. An extra meal is appreciated and should be given at midday to supplement the cat's normal breakfast and supper. It is better to add a meal rather than increase the size of the usual meals for as the uterine horns enlarge, the stomach area may be restricted. The additional meal should be of good quality, high protein food, for it is vital to maintain fitness, not allowing the queen to accumulate extra fat or to lose any muscle tone.

Week Six

During the sixth week, the queen looks pregnant even to the casual observer and may seem less inclined to exercise herself. She should be encouraged to keep fit by taking walks and chasing after a trailed string or feather. The kittens are growing rapidly at this stage, but it is virtually impossible for even an experienced veterinary surgeon to feel individual embryos which have now become larger, softer and elongated, as opposed to the tense ball-like masses of early pregnancy. This rapid growth rate drains the queen's calcium reserves and therefore it is important to provide her with this essential mineral in an easily assimilated form. Some queens enjoy milk without any digestive upsets, and can take all the calcium they need by drinking fresh cow or goat milk. Some cats find diluted evaporated milk more acceptable, while others, notably the Siamese-derived varieties, may lack the enzymes necessary to digest milk products, and need calcium added to their meals by means of proprietary tablets or powder.

Week Seven

Perhaps the most exciting week of pregnancy is the seventh, during which the kittens may be seen and felt moving within the cat's abdomen. This is known as 'quickening' and the queen seems to enjoy the new sensation, often stretching her body sinuously along the floor, then rolling over from side to side. The quickening sensation seems to trigger the nesting instinct too, and the queen starts to explore possible places for the birth, looking into every drawer, cupboard and dim corner, including areas, which to the human, appear totally unsatisfactory. As soon as this urgent searching begins, the breeder should provide one or more kittening boxes placed in convenient places in the home.

Kittening Boxes

Some breeders have special outdoor, heated accommodation in which their queens can produce and raise their litters, but most cat lovers prefer the enjoyment of having litters delivered and reared in the comfort of the home. Special kittening boxes are available complete with integral heating elements, or with overhead lamps, but it is generally better to use disposable materials for parturition, so that there can be no risk of the gradual build-up of low-grade infections during subsequent births.

The queen needs somewhere dark, warm and safe, and the box should be large enough to allow her to stretch out in order comfortably to nurse her litter, but also small enough to enable her to use its sides to push against with her hind legs as she strains in labour.

A large and rigid cardboard box makes an ideal kittening environment, but care should be taken to check its previous contents. Avoid cartons used for packing detergents, soaps or disinfectants, or any other substances which could be toxic, harmful or irritant to the queen. Cut a round or oval hole in one side with a sharp serrated knife with its lowest point about 15 cm (6 in.) from the bottom of the box. This makes the interior draught free, allows the queen to get in and out without undue effort, and will prevent the young kittens from tumbling through the opening when they first start to toddle. Put a thick layer of clean newspapers in the base of the box, having first wiped its interior with a sponge dipped in a suitable disinfectant. Put several sheets of kitchen paper towel on top, and set the box on another thick layer of newspaper to insulate its base. The box should be sited in a convenient spot, where

you can attend during the birth and during the kittens' first three weeks of life.

Special electric heating pads are available for setting in the base of such boxes and provide added comfort. These pads induce the queen to choose your box for her nest rather than all the other areas she may have inspected. The open top of the box should have a removable lid, easily made from a flat piece of cardboard or hardboard. This keeps the interior suitably dim, but allows easy access for attending the birth. The negligible cost of such a kittening box enables it to be disposed of and replaced as necessary, ensuring complete hygiene.

Weeks Eight and Nine

During the last fortnight of her pregnancy the queen should be encouraged to go into the kittening box from time to time, and should be firmly discouraged from using the rest of the house. If the heating pad or lamp is switched on, she will enjoy going into the warmth and seclusion to rest and relax. She should be given four small, well-balanced meals each day and unlimited supplies of freshly drawn water. She may be so bulky that she is unable to self-groom properly, and so she should be gently brushed and combed daily to removed loose hairs. It may be noticed now that a considerable quantity of hair is shed, and in nature, this would provide a warm lining for the nest. In dark-coated breeds, flakes of dandruff may be noticed in the coat, all part of the pre-birth moult. If the queen is carrying a large litter it is probable that she will be unable to clean her anal area, and this should be gently sponged each day using warm water and a little soft soap, rinsing with clean water and drying carefully. Any soreness in this region should be treated with an application of vegetable oil or petroleum jelly.

The nipples should be checked, and if dry or cracked, they too can be massaged with vegetable oil. In longhaired cats, the fur should be trimmed away from the nipples, using blunt tipped scissors and any soiled hair may be clipped away from the genital region. The queen's ears must be kept spotlessly clean, and her coat should be free from any sort of parasite. Pest powders must never be used on pregnant or nursing queens, but a very fine-toothed comb used regularly will ensure that she does not harbour fleas or lice. The queen's teeth should be examined too, for any infection in her mouth can pass to her newborn kittens when she uses her teeth to sever the umbilical cords and her rough tongue to lick the babies clean. If you notice any sign of infection, consult your veterinary surgeon.

During these two weeks the queen takes extra care of herself and spends a great deal of time meticulously self-grooming her abdomen, paying particular attention to her enlarging breasts. She takes pains to ensure that it is safe to jump up or down, and tests the widths of openings before passing through. She becomes very affectionate and rolls on the floor from time to time to rearrange her kittens. She may become constipated, and if so this should be corrected by diet rather than by medication, adding a little vegetable oil or low-calorie margarine to her evening meal. Encourage her to go into her maternity box for a short while each day, where she will probably spend some time tearing and shredding the neat pile of paper into a soft heap.

During the last few days of pregnancy the queen's instinct to nest strengthens and she may damage clothes and other personal belongings if she is allowed to rummage around in wardrobes, drawers and cupboards. The first signs of the approaching birth are gradual and interesting to observe. The milk may come into the enlarged breasts two or three days before parturition, or it may not appear until the birth takes place. The abdomen does become exceptionally pear-shaped however, and the bulge appears to move back towards the cat's pelvis. The round end of the hip bones on either side of the spine become very prominent as the muscles which lie between them and the tail root soften, and their new elasticity may be gently felt with the forefinger and thumb. Hair loss may increase and the cat may crave affection.

The queen's voice changes dramatically a few days prior to kittening, and is especially noticeable in the more vocal, foreign breeds. Within a few hours of the onset of labour, the queen may refuse all food although she might take liquids. She may go repeatedly to her litter tray and scratch in a dispirited way or she might vomit. All these are signs that labour is commencing, and the cat should be kept calm and quiet, with access to her kittening box.

In order to determine accurately the onset of the first stage of labour in the cat, her temperature should be taken rectally, morning and evening, from about the 61st day of gestation. At the onset of labour, it will be seen to drop markedly from its normal $38°-38.6°C$ ($100.5-101.5°F$) to $36.6°$ or $37.2°C$ ($98°$ or $99°F$) and the kittens are on the way.

KITTENING

What to Have Ready

1 A polythene bag containing small squares of terry towelling which have been sterilized by boiling and allowed to dry in a germ-free atmosphere.

2 A pair of sterilized blunt-ended surgical scissors wrapped in a clean linen guest towel.

3 A bottle of astringent antiseptic lotion obtained from your veterinary surgeon.

4 A roll of cotton wool.

5 A roll of kitchen paper towelling.

6 A rubber or stone hot-water bottle.

7 A disposal bin or bag.

Labour: The First Stage

When a maiden queen experiences her first stage of labour she may pace the floor or show symptoms similar to those of 'calling'. She may visit her toilet tray continuously and be generally restless. She may not go near her maternity box, even when the first kitten is well down in the birth canal. The more experienced queen will generally rest quietly in her box, occasionally tearing up the nest material and relaxing right through this long stage. Some queens discharge from the vagina during the first stage and colostrum may drip from the nipples. Uterine contractions may be noticed as the first kitten slowly moves from its resting place in the uterine horn to the body of the uterus and down to the cervix, and these may be accompanied by a break in the queen's rhythmic breathing pattern. Some queens pant as their pulse rate increases, and some queens may tremble while others might growl fiercely at the discomfort. The first stage may last for 24 hours, especially in maiden queens, but unless a cat is clearly distressed, no human intervention is necessary.

Labour: The Second Stage

When second-stage labour begins, the queen must be persuaded to go into her maternity box if she has not already done so. Some queens want literally to have their paws held by their owners, sitting next to the box with the lid removed, while others get along better if they are left quietly alone. In this case, the breeder should take an occasional look to check that everything is proceeding normally. Fierce contractions ripple along the queen's flanks as the first kitten moves into the birth canal. Such contractions may be spaced apart as much as half an hour or so, or may be every few minutes, increasing in regularity until just before delivery when they occur every 30 seconds. The queen may lie on her side, levering her feet against the walls of the box, or she may lie on her chest, pushing her head against the box. A few queens stand and squat in order to gain more leverage in expelling the kitten. Sometimes, just prior to the expulsion of the first kitten, an empty sac of fluid may be passed which often causes great concern to the novice breeder; however, this sac is quite normal and helps to prepare the birth canal for the passage of the kittens.

The Birth Process

The first kitten may be presented head first or tail first. The head-first presentation is considered to be normal, but most breeders will agree that their litters are born with equal numbers of head-first and tail-first kittens and the latter rarely cause problems. The kitten is enclosed within a protective membrane and is surrounded by straw-coloured or greenish fluid before birth. Sometimes it is delivered completely enclosed in this sac, or the sac may rupture during delivery, causing the fluid to discharge from the queen's vulva. In a perfect delivery, the sac, complete with kitten, will be delivered in about 20 minutes. The kitten im-

mediately flexes its neck, and helped by the rough tongue of its mother, frees its head from the membranes and takes its first gulp of air. The queen continues to lick and clean the foetal membranes from the kitten's head and body, stimulating its breathing reflex and being rewarded by squeaks and cries. From the kitten's navel, the umbilical cord extends back through the queen's vulva to the placenta, which a small contraction causes to be passed after a few minutes. The queen should naturally sever the cord in cleaning the kitten, swallowing all the membranes shreds as she licks them from the tiny body, then devouring the placenta or afterbirth, which resembles a piece of liver. She generally eats along the cord from the placenta end, stopping at a point about 2.5 cm (1 in.) from the kitten's body. This cleansing is a natural function, and though it might seem distasteful to some owners, it should not be discouraged. The contents of the placenta are rich in nutrients and hormones, and as well as sustaining the queen during and after the birth, are thought to help stimulate her milk supply and to encourage the eventual shrinking of the uterus.

NATURAL BIRTH: 1 The kitten is born within a foetal sac

2 The queen tears off the foetal membrane

5 The queen licks the kitten dry

3 The kitten gasps, taking its first breath

4 The queen severs the umbilical cord

6 Queen and kitten at rest

As the queen cleans her first kitten, strong contractions may occur again, as the next kitten passes along the birth canal.

Cats vary considerably in their birthing patterns. Some have their kittens at regularly spaced intervals, some have one or two kittens, then take a long rest before continuing their labour and it has been known for queens to have their first kitten one day, and a second kitten 24 hours later. Unless the queen strains for a long time, or seems to fluctuate between one stage of labour and the other, it is safe to assume that things are proceeding normally.

A litter usually consists of three to five kittens, although some queens may have a solitary youngster, and others produce very large litters of eight, nine or ten. When the queen has delivered her last kitten, she settles down, curls her body around the litter and encourages them to nurse, licking them and nudging them towards her swollen breasts.

Helping with the Birth

It must be emphasized that human intervention in the kittening process is only necessary when things begin to go wrong, or if the kittens are born so quickly that the queen is unable to cope. Many breeders interfere, helping to deliver each kitten and taking over the cleaning from the mother cat. They do more harm than good and statistics prove that naturally delivered animals fare better than those assisted into the world. The queen's owner should be at hand to comfort and observe, helping only in an emergency. Some queens, particularly those of Siamese and related varieties literally insist on their owner's presence, and if left, will climb out of the

A normal breech birth

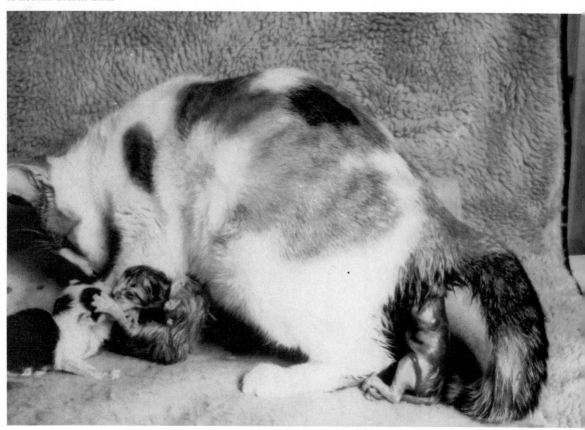

maternity box and go to find them, even with a kitten in the presented position.

Delivering the Half-presented Kitten

If a kitten remains half presented for some time, and despite constant straining by the queen, appears to be stuck, it is quite in order to lend a helping hand. A kitten complete within its sac is often born without trouble, sliding into the world like a well oiled capsule. When the sac has ruptured however, the fluid has been expelled, and the kitten begins to dry out, causing considerable drag. When half the kitten is outside, a small square of terry towel may be placed over the exposed part, enabling a firm but gentle grip to be placed on the tiny slippery body. As the queen strains with the next contraction, the kitten should be eased slowly and steadily in a downward curved movement towards the queen's belly, until free.

Be careful not to pull hard, and make sure that the curved angle of traction is correct so as not to hurt the queen, timing the tension to coincide with the contractions. The queen may cry out as you ease the kitten free, then will rest. Do not pull on the umbilical cord until the next contraction, which may not take place for some time. Try to draw the kitten gently towards the queen's head, taking care that no pull is placed on the umbilicus or a hernia could result. If the queen shows no inclination to clean the kitten, use a piece of kitchen tissue to wipe the mouth and nostrils free from mucus and to clear the foetal membrane from the head. It is important that the kitten breathes before the cord is broken or cut. With luck, the placenta will be passed easily and smoothly, and it should be placed near the queen's head so that she may be encouraged to eat it and deal with the umbilical cord. Sometimes, placing the cord across one of her forepaws helps to stimulate the cleaning instinct. If she ignores the kitten, its placenta and cord, human aid is necessary. Having cleaned the kitten's face, hold it gently, head down, and wipe its body fairly firmly from tail to head, in an action intended to simulate the licking of the mother's rough tongue. Use a piece of terry towel, and work until the kitten is dry. This action also serves to clear any amniotic fluid from the breathing passages.

ASSISTED BIRTH:

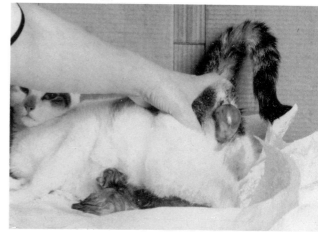

I Helping as the queen strains to pass the head

2 The kitten's body is freed

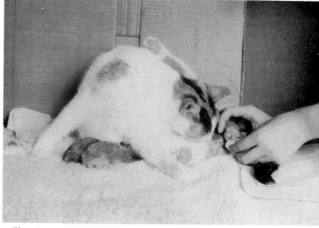

3 Cleaning mucus from the nose and mouth

4 Stimulating circulation

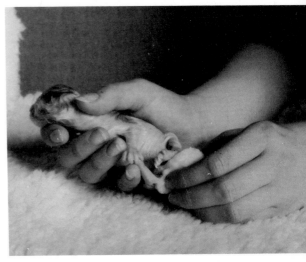

6 The cut end of the cord is pinched

5 The cord is cut 2.5 cm (1 in.) from the kitten's body

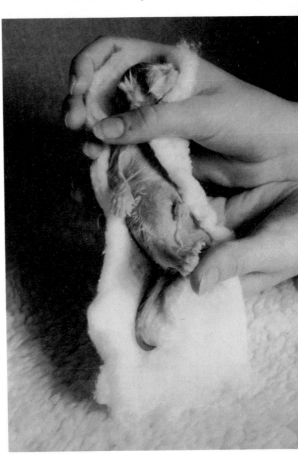

7 The cut and sealed-off end of the cord

Dealing with the Cord

There is no hurry to break or cut the umbilical cord, and it is best to wait as long as possible, in case the cat decides to take over. To sever the cord, round-tipped scissors with fairly blunt blades may be used. They should be sterilized and used to cut the cord about 2.5 cm (1 in.) from the kitten's body. Wash your hands and pinch the cord between the finger and thumb, then cut close to your finger tip, half to one inch from the kitten's naval. The blunt scissors help to seal the cut end, preventing bleeding, but if sharp scissors are used, maintain your grip on the stump for a moment or two. A little astringent antiseptic lotion may be applied to the cut stump to prevent infection. Within a few days the cord dries up and shrivels, and falls off, leaving a neat navel. Experienced breeders can sever the cord with thumbs and forefingers only, pinching it and drawing it apart at the natural severance point but it is important to be shown exactly how to do this by a veterinary surgeon or expert feline midwife, otherwise it might result in a little umbilical hernia. If the mother cat is busy producing the next kitten, her first offspring may be placed on a well wrapped hot water bottle next to the queen until she is ready to accept it. Even if the mother cat refuses to clean the first kitten, she should be encouraged to deal with each subsequent kitten as it arrives, or she may follow exactly the same pattern at subsequent birthings. Persistent human interference can result in a queen becoming quite helpless as a mother.

Difficult Births

Although most litters are born normally and quite naturally, a breeder should always be prepared for things to go amiss. When the labour has continued for longer than usual, or the queen becomes exhausted, or conversely starts to throw herself down in her box, it is vital to call for veterinary aid. In most cases the vet will ask for the queen to be taken to his surgery, a safety measure in case she needs surgical assistance. If possible, take her in her maternity box; failing this, put her in a safe carrier and comfort her during the journey while someone else drives the car. If the queen displays uterine inertia, the veterinary surgeon may inject a sub-

stance to help her contract and expel the first kitten, for if a kitten is delayed for too long in the birth canal, it may die. There is a limit to the length of time that the tiny animal can survive before birth, after the placenta has become detached from the wall of the uterus. The movements of live kittens help the queen to bear down, but a kitten which dies either before the onset of labour or during birth often causes an obstruction.

The veterinary surgeon might find that the kitten is badly presented. The best presentation is one in which the kitten comes head first with its forepaws neatly tucked up under its chin so that it merely has to dive forwards into the world. In the normal tail-first presentation, the kitten's hind legs and tail come first, followed easily by its smooth body, then another contraction generally eases the shoulders and head into the world. Trouble occurs when the kitten is turned so that its head is trapped near the cervix and the back of the head and shoulders block the entrance to the vagina. In this instance the vet may carefully manipulate the kitten, pushing it backwards, disengaging the trapped head, and drawing it forward again into the birth passage.

Butt Presentation

The kitten presented with its hindquarters foremost and with its hindlegs extended towards the head can also pose problems, and if the kitten is very large the queen cannot pass it, no matter how hard she strains. Again the vet will try manual adjustment of the kitten's position, but bearing in mind the small internal dimensions of the cat, it will be readily understood that this is a delicate, skilled procedure, and one which should never be attempted by a novice.

Caesarean Section

Some abnormally presented kittens may be delivered safely by the veterinary surgeon with the aid of special forceps, but others sometimes die in the long, often arduous process. When there is no alternative, the vet may perform a Caesarean section to remove the kittens. Even when expertly carried out, this operation may make future litters difficult if not impossible to produce normally due to the formation of adhesions. Such adhesions change the shape of the uterine horns and may complicate the passage of kittens into the vagina. Most veterinary surgeons perform a Caesarean section only as a last resort, and often ask permission to spay the cat at the same time. Great care is taken in the selection of the anaesthetic used, so that the kittens are virtually unaffected, and the operation is performed in safe, sterile conditions at the surgery, while the anxious breeder nervously waits in an ante-room. When the kittens are delivered, the vet will probably have them passed over to the breeder for stimulation, cleaning and drying, while he gets on with the job of suturing the queen's abdominal incision. The vet may allow the queen to go home as soon as she recovers from the anaesthetic, or he may decide to keep her under observation for a few hours. The kittens may be kept with the queen so that she accepts them as she recovers consciousness, or the vet might prefer them to go home with the breeder.

In her bemused state, coming out of the anaesthetic, the queen could bite or kill the babies, especially as they scrabble for milk near her sore, stitched abdomen. If they are taken home, they should be kept at a temperature of $24°-26°C$ ($78°-80°F$) and rubbed quite dry and clean. They do not need feeding for several hours as long as they are warm and kept in a quiet dark place, in fact syringe feeding can often do more harm than good. When the queen returns, she must be carefully introduced to her litter, and will probably have been given an injection to stimulate her milk supply, and help to induce maternal feelings. Some queens take several days fully to accept their kittens born by Caesarean section, and great patience is needed on the part of the breeder. If the queen has plenty of milk, she should be gently but firmly held while the kittens suck, and this should be done every two or three hours. The queen should be petted and encouraged all the time and should soon show some interest in the kittens. The stitches holding the incision closed are generally removed after seven to ten days.

Abnormal Kittens

Not every kitten that is conceived is destined to become an adult cat. Some kittens die in embryo for a variety of reasons, some may die during birth and some soon after birth. A small percentage of kittens are born deformed in some way, and may have to be destroyed. You should not attempt to rear a kitten born with a serious defect, for it can never grow to become anything but a freak.

Perhaps the most common deformity in some strains of pedigree cats is known as cleft palate, where the bones across the roof of the mouth have failed to join together during the process of gastrolation, before the kitten's birth. A kitten with this defect is born normally and seems quite content for the first few hours, then becomes increasingly restless and noisy, nosing its way from one nipple to another in its quest for nourishment. If watched closely, milk may be seen dripping from the kitten's nostrils as it tries to nurse, and a slight wheezing sound is heard. There are varying degrees of the extent of the cleft in the palate. In some cases the split runs about halfway along the roof of the mouth and though the kitten can suck it cannot obtain enough milk for its needs and some seeps into the nasal passages. In some kittens, the cleft may be so extensive that the top lip is also involved and the animal cannot even suck. If the cleft is minor, a kitten may survive with extra care and hand feeding but should be neutered at maturity so that it cannot pass on this defect to any offspring.

Sometimes, kittens are born with legs so twisted that the hindlimbs seem to have been attached the wrong way round. It is important for such kittens to be examined by an experienced veterinary surgeon to determine whether or not all the bones are present in the limbs, and that the hips are neither dislocated nor deformed. If the bone structure is complete, the twisted condition is caused by slackness of the muscles, and may have been caused by a bad position prior to birth. Gentle manipulation and exercise can provide sufficient physiotherapy to correct the legs, and produce complete recovery, but again, it would be better for such kittens to spend their adult lives as neutered pets rather than breeding stock.

At first the twisted-leg kitten drags its body forward, scrabbling with the hind legs, and the fur on the stifle joints and over the toes may become denuded. It is very important to prevent injury to this region by providing plenty of soft bedding at this stage. As the kitten gets stronger, it must be helped by encouraging it to use the hind legs. These can be exercised several times daily, flexing and stretching the joints and letting the little animal kick against the palm of your hand. Make sure that the kitten gains weight at the proper rate, and if the veterinary surgeon agrees, add extra calcium and vitamins to its diet.

If the hind legs are held very wide apart, a small pair of boxer-shorts can be sewn from crepe bandage and put on the kitten for play periods each day. These help to support the muscles around the hip joints and encourage the youngster to use its hindlimbs in the correct fashion.

Very occasionally a breeder may be distressed to find that a kitten is born with its eyes open, instead of being tightly sealed shut in the normal manner. Until quite recently it was the considered opinion of the experts that such kittens would be blind, and so they were destroyed soon after birth. Some breeders however, decided to rear their open-eyed youngsters, and it was found that most of these developed normally and proved to have perfect eyesight. The important factor in their successful rearing seemed to be the provision of very dim light conditions for the first ten days of the kitten's life, corresponding with the length of time that the eyelids might be expected to remain closed, although some breeds, notably the Siamese varieties, do tend to open their eyes from the third day onwards. In some cases it has been found that the open-eyed kittens in fact close their lids naturally after a day or so, to re-open them along with their normal littermates at the natural eye-opening period. Some kittens with eyes open at birth have missing or deformed eyelids, and as these are usually allied to other internal deformities, such kittens usually fade away shortly after birth. If not, veterinary advice should be sought as to whether or not the kitten should be put down.

Perhaps the most distressing of all birth deformities is that of the umbilical hernia which is so extreme that the kitten's intestines are enclosed in a pouch of skin outside its body cavity. Such kittens are often born intact and are strong, crawling eagerly towards the queen's nipples. Sometimes the queen in severing the umbilical cord, ruptures the skin causing the intestines to spill out. Such deformed kittens must be put down immediately, for there is absolutely nothing that can be done for them and as the condition is probably hereditary, the queen's suitability for further breeding should be carefully considered.

Some minor deformities occur in kittens, but do not affect them adversely. One of these is polydactylism, which indicates the presence of extra toes. This may vary from one extra toe on one forefoot, to the extreme where the cat may look as though it has two complete paws on each leg. As this characteristic is produced by a dominant gene, it means that a kitten must have one affected parent, and if all polydactyl cats were neutered, the condition would disappear.

Bobbed, bent or kinked tails are quite common in cats, especially in those of Eastern origin, but due to careful selection of breeding stock over the years, only an occasional slight kink is seen in British-bred pedigree felines of today. Some of the first Siamese cats imported into Britain had short, knotted tails, and to this day, the Siamese and its derivatives are still more prone to the deformity than other breeds.

Some deformities have been valued and cats showing them have been used to start new breeds. The Manx is a cat in which the tail is so reduced that it forms either a tiny stump or is totally missing. The Scottish Fold is a cat of British type, with tightly down turned ears and the Rex cats have genetically deformed coats, causing them to look curled.

6

Showing and the Cat Fancy

BRITISH CAT SHOWS

A large number of cat shows are held each year at various venues throughout the country. Showing can be an absorbing hobby and enables you to make many new friends, and the well run show often provides a pleasant day out for the entire family. If you want to breed a successful line of your own cats, the cat show provides a parade of male cats as potential studs for consideration and gives you the opportunity to meet with established breeders of the variety you fancy.

Showing your own cat gives you the opportunity to have its type and potential assessed by experts during judging and you soon see how your cat compares with others of the same breed. In Britain kittens may be shown when aged from three to nine calendar months, and are considered adult at nine calendar months. Dates are taken as being effective on the actual day of the show, so a kitten to be entered for competition on 1 October, for example, must have a birth date from 2 January up to and including 1 July, while a cat born on or before 1 January would be entered in the adult section. The showing of neutered male and female cats is actively encouraged in the Cat Fancy and many special classes are offered for such cats.

In Britain pedigree cats and kittens may be registered with the Governing Council of the Cat Fancy, usually referred to as the G.C.C.F. This body is responsible for the authorization and licensing of cat shows throughout the United Kingdom and

produces a list of the current season's managers. A copy of the official list may be obtained by sending a cheque or postal order for £1 (1982) to the Secretary, Mrs W. Davis, G.C.C.F., Dovefield, Petworth Road, Witley, Godalming, Surrey GU8 5QW. Further information is then obtained by sending a self-addressed envelope to each of the managers of the shows you select. Ensure that these envelopes are at least 23 cm × 15 cm (9 in. × 6 in.) because they will be used for sending bulky schedules, sets of rules and official entry forms. Show schedules are generally available three to four months prior to the show date, and the closing date for entries may be at least two calendar months before the show, to enable the official catalogue to be compiled and printed. It is necessary to plan well ahead if you intend to embark on a show career. Most shows have special sections for non-pedigree cats and if your pet is of unknown ancestry, ask for a Household Pet or Non-Pedigree Section entry form when applying for your show schedule.

Of the three types of cat show, the most important is the Championship, at which special Championship Challenge certificates may be awarded to adult cats winning their open or breed classes. Neutered cats winning their open classes may be awarded similar Premier Challenge certificates. A cat must win three such certificates at three separate shows, and awarded by three different judges, to achieve full championship status, while a neutered

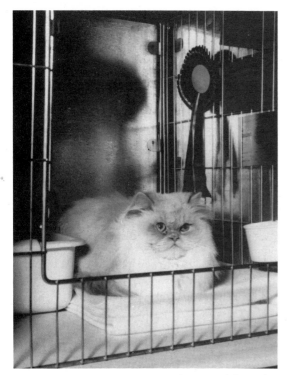

G.C.C.F. show Championship Challenge and Best of Breed winner

cat gaining the same number of wins becomes a Premier. Champion cats and Premier neuters are entitled to affix their titles to their names and may continue to gain further honours in Grand Champion and Grand Premier classes.

Sanction shows, with one special exception, do not permit the award of Challenge certificates, but are considered as dress rehearsals for the larger championship events. The one exception is the unique show put on annually by the Kensington Kitten and Neuter Cat Club which has sanction status because, although no entire adult cats may be entered, the Club is permitted to award Premier Challenge certificates to its neuter winners.

Exemption shows are for newly established clubs and are friendly affairs complying with G.C.C.F. regulations, which have special rules and classes of their own.

When your stamped addressed envelope is returned to you by the show manager, it will contain a show schedule, entry form, show rules and possibly

a leaflet with information about advertising in the official show catalogue and other ways in which you may wish to support the show fund. The schedule gives full details of the club running the show, a list of its officers, officiating judges and vets, the date and venue, and the name and address of the show manager or assistant to whom the completed entry forms must be mailed.

Open classes, the most important classes of all, are often divided into colours within breeds. The winners of such classes may be eligible for Best of Breed awards at some shows, and some shows offer the still higher award of Best in Show. Miscellaneous classes vary considerably from show to show and you may enter any for which your cat is eligible. Such classes are generally grouped thus:

Kitten	Young cat three to nine months
Adolescent	Cat between nine and fifteen months
Junior	Cat between nine months and two years
Senior	Cat over two years
Veteran	Cat over seven years
Novice	Exhibit not having won first prize
Special Limit	Exhibit not having won more than two first prizes
Limit	Exhibit not having won more than four first prizes
Debutante	Exhibit making their showing debut
Breeders	Exhibit bred by owner
Novice Exhibitor	Must be exhibitor's very first show

Some clubs offer special Charity classes at their shows, and the proceeds are given to a designated charity. By entering such classes your cat is helping felines less fortunate than itself. All the qualifications for entering miscellaneous classes are clearly listed in the schedule, so read it carefully. Club classes are sometimes offered by other clubs wishing to support the show and winners receive special rosettes. You may not enter a class unless your membership of that particular club has been approved and your subscriptions are up to date.

Having read the schedule right through and

noted the rules, read the entry form too, for it is an official document and must be treated as such. Be sure to have your pedigree cat or kitten registration or transfer form in front of you before you begin, as it is essential that the details you enter correspond with the exhibit's official G.C.C.F. records. If you have recently acquired a kitten, be sure that the transfer application is sent to the G.C.C.F. Recorder to arrive at least three weeks before the show date. Use ink or a ball-point pen to complete the form and write clearly in block letters. Enter the registered name of your cat, not its pet name. The sire is the father of your cat, and the dam its mother. The date of birth must be added and the registration number. If this has not been returned to you before completing the form, put 'r.a.f.' in this box, denoting 'registration applied for'. The breeder is the registered owner of the dam at the time of your kitten's birth. Put the open class number first, then follow with the other class numbers and try to keep them in running numerical order to assist with the processing of your entry in the show manager's office. Work out the correct fees including the benching, which is the cost of hiring your cat's show pen for the day, a separate fee for each class entered, and any extra passes that you might require.

Make out a cheque or postal order for the correct amount and attach it securely to the entry form. Read the declaration printed on the form, sign it, and if you own the exhibit jointly with a husband, wife or friend, obtain a second signature. Add your full address and a telephone number so that the show manager can contact you quickly with any query. For an acknowledgement of your entry, enclose a self-addressed, stamped postcard. An envelope for the mailing of your pre-show tickets and tallies from the show manager is always appreciated. A tally is a small disc numbered to correspond with your cat's show pen.

If your exhibit is an unregistered, non-pedigree pet, you may only enter in a special section of the show and you need a different entry form. Pedigree or pet, be sure to send in your entry well before the published closing date, for showing cats is a popular hobby and many shows are oversubscribed. Late entries, even before closing date, may have to be returned.

Preparation for Showing

The condition of a cat is built up from inside, so a carefully formulated, well-balanced diet is the first essential and might perhaps be preceded by a course of worming treatments. Do not give your pet double doses of vitamin pills, but aim to build up the muscles and general fitness. A slightly increased meat ration, combined with exercise over a period of four to six weeks, and ten minutes' extra grooming each day usually helps to improve the look of any cat.

Longhaired cats need regular attention and show preparation is important: in addition to usual grooming it may be necessary to remove grease in the coat by giving the cat a bath. After bathing, the thoroughly dried coat is combed right through. Powder is often used to clean and separate the hairs in the Persian's coat, but any powder left in the pelt when the cat is on show could end in disqualification. Shorthaired cats may be bathed if necessary but often look immaculate after a daily polish with a soft dry chamois leather, or a thick pad of warmed cotton wool.

All cats need to be carefully prepared for show day, and special attention should be paid to the greasy area on the top of the tail near to its root, and to the claw beds which can get very dirty. The ears must be spotlessly clean too, as must the inner corners of the eyes and the nostrils.

Show Equipment

For use in the show pen you will need a plain white blanket, which must not be of a cellular weave or bound with ribbon. The toilet tray, water dish and food bowl should all be of unadorned white plastic. White tape, ribbon or narrow elastic is necessary for fixing the numbered disc or tally round your cat's neck. A safe carrier in which to transport your cat to the show is also necessary, for exhibits are not accepted on leads, or carried in their owners' arms.

It is useful to take a small show kit including a small sponge and a suitable diluted disinfectant for cleaning the show pen; the cat's grooming equipment; his food and water; a small pair of scissors; a pencil or pen for marking results; a small bag for rubbish and some very comfortable shoes for yourself. Be sure to pack all the show documents and the cat's vaccination certificate.

When the great day arrives, be sure to get up in time to make an early start on the journey to the show hall. Your cat should have been groomed to perfection and brought into top condition by this time and your show bag should have been packed the previous evening. Feed the cat early, give him time to use his litter tray before packing him securely in his carrier, then set off. Plan to arrive at about 8 a.m., which is the usual time for vetting-in to commence.

If you have received your show documents by post, have them ready. Otherwise they will be handed to you at a desk as you enter the foyer. You will be directed to the veterinary area by show officials. When your turn comes, the vet's steward will take your cat from its carrier and the veterinary surgeon will check its general condition. He is looking for indications of infection such as watery eyes, discharging nostrils, a sore throat, or staining under the tail caused by diarrhoea. The coat is checked for fleas or other parasites, or lesions caused by the ringworm fungus. The ears are checked for earmites. Male cats are examined to ensure that they are entire, and females to confirm that they are not pregnant. You should have already assured yourself that your cat is in good health, so your vetting-in slip will be signed, and you may pass into the hall.

Quinteiro Hot Chocolate – Best Burmese at the Jersey Club show, 1981

Inside the show hall rows of steel pens are arranged in numerical order, and you must find the one which corresponds to the number on your cat's tally. White-overalled officials will be pleased to help if asked. You may want to clean your pen with disinfectant before arranging the litter tray. Reassure your cat, and put him gently in the pen. If he uses the toilet tray, empty it into your refuse bag and refill with clean litter. Groom your cat if necessary and settle him down on the folded white blanket, tying the tally around his neck so that it fits snugly but is not too tight.

You may feed your cat if you wish but the dish must be removed before judging commences. Be sure to provide a white bowl half filled with fresh drinking water. If you have had problems in purchasing the correct show equipment, you should have time to buy it from one of the stands specializing in cat accessories, situated in the hall. Around 10 a.m. an announcement is made, asking exhibitors to 'clear the hall'. This means that judging is about to commence and you must leave the hall until the open classes have been completed. Check that your cat has settled, take out the food dish and conceal it with your other belongings behind the drapes, underneath the pen. Leave the hall quickly and buy yourself a copy of the show catalogue, on sale only after the commencement of judging. Check through the catalogue – if there appear to be some errors in your entry, report them to the show manager immediately so that they can be corrected without delay.

Between 12 and 1 p.m. you will be allowed back into the hall and may feed your cat if you wish. Judging continues through the afternoon and you should stand well away from your cat if you see a judge and steward approaching. You must not speak to a judge until the end of the day, when all the classes are completed, for everything is anonymous and very fair, ensuring that a novice with a good cat is as likely to succeed as an established fancier. Each judge notes awards on numbered slips in a judging book and a copy is posted on a special board. Check your results by looking at the board for the members of the classes in which you have entered, then check down the judging slips until you find the number of your cat's pen, against

which, if you are lucky, an award will be entered!

Prizes are indicated by 1, 2 or 3 for first, second and third. R stands for reserve, the fourth position, and VHC means very highly commended. HC is highly commended and C is commended. These last three awards are given to pleasing exhibits outside the first four or five placings. In open classes, adults may also have the letters CC, meaning that a Champion Challenge certificate has been won, while neuters may have PR, indicating the winning of a Premier Challenge certificate. BOB stands for Best of Breed.

If your cat has won an award you may be entitled to prize money, which is collected from a pay-out point in the hall, or you may received a rosette which will be affixed to your pen during the afternoon.

Whether your cat wins or loses you are sure to have a happy day, and at about 5.30 p.m., when the show closes, you must gather up your cat, its belongings and prize cards, and leaving the pen area as clean and tidy as you found it, set off for home.

AMERICAN CAT SHOWS

In the United States show procedures and regulations vary considerably. Shows generally run right through the weekend. The cats are penned in groups in the main body of the show hall and their owners stay with them at all times. Most of the pens are decorated with colourful drapes, ribbons and photographs and the onus is on each owner to carry his cat to the relevant judging ring as each class is announced. In each judging ring the judge presides and is assisted by two or three clerks. The cats are placed in a row of empty pens immediately behind the judge who takes each in turn and placing it on the table assesses its good and bad points aloud to an enthralled audience. Having examined each cat in the class the judge places prize ribbons on the cages of the lucky winners who are then taken back to their own decorated pens.

American judges are highly trained and fully qualified, and are paid a small fee for their work. They may handle 200 or more cats in each show and finish the second day by placing their top exibits in order for Finals Awards.

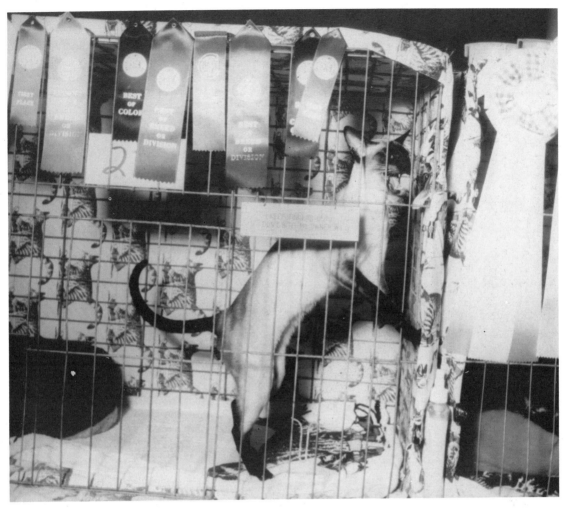

Each of the leading Associations produces a list of show dates, which may be obtained by writing to the relevant addresses (see page 98).

Cat Fanciers' Association show – a Seal-point Siamese exhibit protests at being penned

THE GOVERNING COUNCIL OF THE CAT FANCY

In 1871 Harrison Weir, artist, writer and felinophile, helped to promote interest in the breeding and showing of pedigree cats by organizing the first cat shows at the Crystal Palace in London. In 1887, at a four-day show held at Alexandra Palace, a dedicated group of exhibitors decided to form the National Cat Club with Harrison Weir as the first elected president, and the British Cat Fancy was born.

For the next 23 years the Club carried out all the functions of registration, instituted championship awards and formulated the detailed rules and regulations to govern the running of cat shows. The Club's first official list of Cats at Stud was published in 1895.

A friendly rivalry was established in 1894 when the Scottish cat fanciers formed themselves into a group and held cat shows in Glasgow. Several other clubs and societies were formed and all agreed to run their shows under the rules of the National Cat Club. However, the cat world's peaceful and pleasant atmosphere was shattered in 1898, when the Cat Club was formed under the presidency of Lady Marcus Beresford, and functioned with its own rules and licensed shows. Breeders and exhibitors of the day soon found themselves split into two rival factions, and in an impossible administrative situation.

After several traumatic and troubled years, a meeting of all cat clubs was called in Westminster, London, on 8 March 1910. Nineteen delegates were present and the meeting was chaired by Russell Biggs of the National Cat Club. A decision was made to form a new body to be called the Governing Council of the Cat Fancy, and the National Cat Club graciously handed over all its powers to the new Council, enabling one register and one set of rules to be followed by the whole Cat Fancy.

Today, clubs and societies with sufficient subscribers may become affiliated to the G.C.C.F. and are then privileged to send delegates to council meetings. The Council consists of representatives of more than 60 all-breed clubs and specialist clubs, and the larger clubs are entitled to more than one delegate, each of whom has voting powers.

The constitution of the G.C.C.F. defines its obligations to member clubs, and controls all matters relating to the registration and exhibition of pedigree cats throughout the United Kingdom. The G.C.C.F. has an annually elected chairman, vice-chairman, and treasurer, and the other members of council are delegates democratically elected by their clubs. A full-time secretary is employed and a team of registrars deal with the recording of kitten births, official names and the transfer of ownership of registered stock. Specialist committees exist within the Council to deal with finance, discipline of rule breakers, genetics and cat care. The everyday workings of the Council are managed by an executive committee of 16 annually-elected delegates and a legal adviser is always available.

Since its conception, the constitution of the G.C.C.F. has occasionally changed in some respects, but although its rulings are strict and exhibition and registration regulations must be obeyed to the letter, the fancier usually feels that he receives just and compassionate treatment from the Council in all feline matters at all times.

THE AMERICAN CAT FANCY

Unlike exhibitors in Britain, those in the United States of America have a variety of registration bodies from which to choose.

American Cat Association (A.C.A.) is America's oldest cat registry, and has been active since 1899. Although it is one of the smaller associations, it has shows in the south-east and south-west regions of the country. Contact: Ms Susie Page, Secretary, A.C.A., 10065 Foothill Boulevard, Lake View Terrace, California.

American Cat Council (A.C.C.) is another small association and is centred in the south-west. A.C.C. has held several shows in a modified 'British' style, excluding exhibitors during judging. Contact: Althea A. Frahm, P.O. Box 662, Pasadena, California 91102.

American Cat Fanciers' Association (A.C.F.A.) is one of three international associations which have affiliated clubs in the USA, Canada and Japan. A.C.F.A. is a very democratic association and offers individual membership as well as club charters and produces a lively monthly newsletter. Contact: Ed Rugenstein, General Manager, P.O. Box 203, Point Lookout, Missouri 65726.

Canadian Cat Association (C.C.A.) is the only all-Canadian registry with activities located mainly in eastern Canada. Its newsletter is published quarterly in English and French. Contact: C.C.A., General Office, 14 Nelson Street West, Suite 5, Brampton, Ontario, Canada L6X 1B7.

Cat Fanciers' Association (C.F.A.), the largest of the American registries, is operated by a board of directors, and changes are brought about through votes of delegates, each club electing one delegate. There is a C.F.A. show somewhere in the U.S.A. almost every weekend, and the Association has affiliates in Canada and Japan. C.F.A. produces a beautiful hardback, annual yearbook full of colour and black and white photographs of American cats of every conceivable shape, size and colour. Contact: Thomas H. Dent, Executive Manager, 1309 Allaire Avenue, Ocean, New Jersey 07712.

Cat Fanciers' Federation (C.F.F.) is a medium-sized association with activities based in the north-east region of the country. Contact: Barbara Haley,

Recorder, 9509 Montgomery Road, Cincinnati, Ohio 45242.

Crown Cat Fanciers' Federation (C.C.F.F.) is one of the smaller associations but it has many shows in the south-east, north-east and in western Canada. Contact: Sister Vincent, Recorder, P.O. Box 34, Nazareth, Kentucky, 40048.

The International Cat Association (T.I.C.A.) is the youngest of the American associations and its friendly atmosphere and innovative methods have resulted in a rapid growth rate. Now the second largest American association, T.I.C.A. has affiliated societies in Canada and Japan. Its club newsletter, *The Trend*, is published six times a year, while the yearbook is a beautiful hardback presentation. Contact: T.I.C.A., 211 East Oliver, Suite 208 Burbank, California 91502. In 1983 T.I.C.A. started an official UK Fancy. Contact: The Registrar, T.I.C.A. (UK), Orchard Cottage, Westcot Lane, Sparsholt, Wantage, Oxfordshire OX12 9PZ.

United Cat Federation (U.C.F.) is centred in the south-west. Contact: David Young, Secretary, 6621 Thornwood Street, San Diego, California 92111.

7

Cattery Care

THE CATTERY

A 'cattery' is the name given to accommodation specifically designed and used for keeping cats. Catteries vary in the size and design and are made to house one or many cats. The cattery may be sited indoors, outside or may combine indoor and outdoor accommodation, depending upon its geographical location. Some catteries are constructed to provide ideal environments for the breeding, rearing or boarding of cats, while some prove to be quite inadequate, difficult to clean and unhygienic.

For the small breeding unit it is usual to have cat houses made of timber, sited out-of-doors and facing south. Timber is warm and comparatively inexpensive and ready-made workshops, sheds and garages of varying sizes can be purchased which need only very small adaptations to convert them into cat houses. Cats only thrive in clean, airy conditions, so the building chosen for conversion should be of adequate size. It is best to build the cattery on a concrete base laid to a slight slope for easy drainage and trowelled to a smooth surface to facilitate cleaning. The timber building should, if possible, be raised on one or two courses of brickwork and the concrete base should extend to provide an exercise area. The timber house should be lined throughout with smooth faced or laminated boarding. Laminated plastic boards used for kitchen construction are ideal and once fitted require no future maintenance. However, the initial outlay can be quite substantial. Smoothly finished hardboard or blockboard is suitable for lining but

must be sealed and then painted with a washable, vinyl paint. A new top coat of this can be applied each year. The ceiling should also be lined to retain warmth at night, and a timber floor is preferable to a floor of poured concrete. Timber flooring should be tiled or covered with vinyl which is taken up the walls to a height of about 15 cm (6 in.) all round and fixed in place with battens or a suitable adhesive. Electric wiring can be run from the house to the cattery through a special exterior conduit made of polypropylene. This can be obtained in all lengths and shapes and has cat-proof fittings. Heating and lighting can then be fitted as required inside the cat house. A cat flap may be fitted in the entrance door and shelves of varying sizes can be screwed to the interior walls to provide sleeping areas and to allow exercise facilities. If the cattery is situated in a cold region of the country, some insulative material should be packed between the lining of the building and the exterior walls.

Outside the cat house a run should be constructed, high enough to allow a human adult to walk around and large enough for the cat to take adequate exercise. The run may be constructed of metal or timber framing and covered with stout wire mesh. The mesh must be fitted to the framing, leaving no gaps, and part of the roof may be covered with opaque or clear corrugated roofing panels to enable the cat to use the run even in very wet weather.

The entrance to the cattery is its most vulnerable

area; gates must be of adequate size to allow easy access but should not be so large that the cats can run past every time they are opened. A gate should be well made with three efficient hinges allowing it to hang well and swing easily. It should be fitted with a spring-loaded clip to hold it shut on the inside and there should be a strong, lockable bolt on the outside. Cat houses used for stud work should ideally have double gates forming a safety area to prevent the possible escape of visiting queens. In larger catteries an ideal arrangement is to have all the cat house runs opening into an enclosed safety passage.

Although grass runs look nice when freshly laid, they can become unkempt and unhealthy. Paved or concreted runs are very easy to clean and can be furnished with pots of grass for the cats to chew on and logs for climbing.

Inside the cattery a set of grooming and cleaning tools may be neatly stored. All cats in a cattery should have comfortable cosy beds, easily cleaned toilet trays and clean non-spill food and water bowls. If it is convenient there should be a shelf or table at a convenient height for examining or grooming the inmates, and a roll of paper towelling, fixed in a special holder high on the wall, is handy

An old cat in a cattery

for wiping up odd spills and accidents. A brush, dustpan and mop can be hung on wall hooks and a small wastebin, with a cat-proof lid, is useful.

Outdoor catteries provide healthy living conditions for cats and can be heated in cold weather by means of infra-red lamps hung over the cats' sleeping areas. Indoor catteries are easier to heat, but even more stringent cleaning methods must be maintained as feline diseases spread more rapidly in enclosed areas than out-of-doors. Indoor catteries are commonly constructed in spare bedrooms, large garages, workshops or basements and the accommodation provided is determined by the number of cats in the colony. There are endless possibilities of layout and design for indoor catteries, but all should have good ventilation, lots of light and plenty of exercise space.

BREEDING ACCOMMODATION

A serious breeder with several queens may decide to construct special accommodation in which the cats can be safely housed away from visiting local toms and possible infections. A cat house or indoor cattery pen may be quite adequate but the breeder may decide to construct special kittening pens and an adventure playground area for growing kittens. A kittening pen should be draught-proof and have some form of easily controlled heating. It is imperative that the whole area can be easily cleaned and sterilized. A queen needs a dark, private place in which to give birth and nurse her family for the first three weeks of their lives. She must be free from stress but have sufficient room to exercise and for her toilet tray and food bowls. Collapsible kittening pens are available and can be useful indoors. The small breeder who allows the cats the run of the house may place a portable kittening pen in a bedroom, living room, study or kitchen and the queen may be let out for long periods of exercise each day. Most breeders have special areas for raising their weanling kittens. From the age of six weeks kittens need the added stimulation of play and a suitable area should be provided. The kittens need plenty of toys and exciting objects within their environment, such as tunnels and caves made from

cardboard boxes. In a large cattery certain areas are generally set aside for the weanling kittens and the mother cats are removed for longer and longer periods each day, until the kittens are totally independent. This special play area must be sited well away from any possible sources of infection, and can be designed to provide an ideal showcase for potential buyers.

BOARDING ACCOMMODATION

The strict regulations for the running of boarding catteries vary from country to country. In Britain these regulations are laid down by licensing authorities and must be observed when designing and constructing the layout and in the general administration of the cattery once it is in operation. A boarding cattery must provide adequate, safe, hygienic accommodation for all its visitors. The inmates must be provided with suitable shelter, bedding and food and have sufficient room in which to exercise. Proper precautions must be taken to prevent the outbreak and spread of disease and all risks of fire should be kept to the very minimum. There must be a responsible person in attendance on the premises at all times.

In temperate climates the ideal boarding catteries consist of outdoor accommodation and generally have small brick or timber chalets with individual, open air runs. In countries which experience extremes of climate, boarding facilities are generally housed inside a large building, artificially heated in the cold months and cooled during warmer periods. All catteries experience the same problems resulting from the regular fast turn-round of visiting cats and certain preventative measures should be taken against serious feline diseases.

Each cat accepted into a boarding cattery should be expected to pass a basic test of health; its temperature should be normal and there should be no sign of discharge from its eyes or nostrils. The ears and coat should be free from parasites and there should be no sign of yellow staining under the tail which could indicate diarrhoea.

Standard routines should be aimed at providing

An ideal boarding cattery, with roomy fresh air runs

hygienic conditions and preventing any chance of cross-infection between cats. Any cat showing even slightly suspicious signs of disease should be examined by the visiting veterinary surgeon and placed under observation in the unit set aside for isolation purposes. All cats should be fed ample amounts of their favourite foods every day and disposable dishes should be used whenever possible. Fresh drinking water must be provided daily in clean, non-spill containers. Each cat must have room in its quarters for its own bed, an adequate toilet tray and possibly a scratching post and its favourite toys. The licensing authorities' requirements set out the minimum standards for

each cat but most boarding cattery proprietors build their units to larger specifications. This means that family groups of two or more cats may be housed happily together. On no account should two or more cats from different homes be expected to share boarding accommodation.

Good catteries insist on seeing current certificates of vaccination before accepting cats into their establishments. Most catteries insist on vaccination against Feline Infectious Enteritis and some also insist on vaccination against the upper respiratory diseases. To cut down the risk of upper respiratory disease being spread by apparently healthy carrier cats, good catteries have solid partitions between each of the boarding units or wide spaces between the runs. Indoor catteries should have efficient extractor fans constantly changing the air.

CATTERY CARE AND MAINTENANCE

Every cattery should be carefully and regularly maintained – it is surprising how much damage and deterioration can occur each year. Outdoor catteries are affected most as heat and cold attack paintwork, wire mesh rusts and distorts, hinges and bolts require regular oiling and water pipes and drains need clearing. Routine maintenance is generally carried out during the slack months; wooden cat chalets or pens are freshly painted or coated with preservative, vinyl flooring is checked and renewed if necessary, cracked glass is replaced, concrete is repointed, roofing is checked and patched, the electric wiring is carefully inspected.

All timber cat houses should be washed to remove algae, and one or two coats of preservative may then be applied. The cats must be removed to other accommodation while the timber is treated and sufficient time must be allowed for the wood to absorb the preservative and for its surface to dry *thoroughly*. Such preservative preparations can be highly toxic to cats and must not be allowed to get onto their coats or the pads of their feet. The timber of the runs should also receive preservative treatment each year and the fixings holding the mesh to the frames must be carefully checked and replaced if necessary. Concrete and paving may be cleared of algae growth by scrubbing with a solution of bleach.

The bolts and hinges on the access gate should be checked carefully, loose screws should be replaced and all moving parts treated with a light application of oil.

Whatever the size of the cattery and whether it is an outdoor or indoor establishment, routine systems of management should be instigated and carefully followed. Catering arrangements are particularly important and animal foodstuffs, dishes and utensils should be stored separately from those of humans. Larger catteries often have kitchens specially designed and adapted for preparing feline food. Store cupboards contain canned foods, com-plete diets and cereals, refrigerators hold fresh foods, such as meat, fish, eggs, milk and cheese. Bulk supplies are often kept in large cabinets or chest freezers. The cattery kitchen has its own pans, dishes, bowls and trays, a stove or hot plate, sink and drainer. Strict attention to hygiene must be observed in the preparation and serving of the cats' meals. Unless disposable dishes are used, feeding bowls must be washed immediately after use, then passed through a sterilizing medium. All water bowls should be washed daily and rinsed thoroughly before being refilled with fresh water.

Most cattery proprietors have their own feeding methods, but the most successful agree that a good variety of quality produce and fresh food produces the most satisfactory results. The best catteries feed individual diets to their boarders, ensuring that homesick animals are always offered their favourite, irresistible meals. The leading brands of complete, dried cat foods have proved a boon for boarding catteries: many visiting cats refuse to eat during the day and their fresh or canned meals cannot be left in the pens as they soon become stale and attract flies. The dried diets, however, remain fresh for many hours, and may be left overnight in the accom-modation, enabling cats to feed at will.

In the boarding cattery the toilet tray from each pen is emptied and refilled with litter, or replaced with a fresh tray if it is soiled. Litter thrown out or spilled on the floor of the pen or cat house should be swept up, the floor covering cleaned with a cloth or sponge dipped in hot water containing a disinfectant known to be safe for cats. Particular attention must be paid to the angles between floor and walls where germs may lurk. Any splashes or marks on the walls should be cleaned, and all surfaces dried with a sponge pad, cloth or kitchen towelling. The cat's bed, fresh toilet tray, food and water bowls are then replaced in the pen or chalet. This may be a good time to check the cat and groom it if necessary.

Fresh disinfectant solution should be used for each cat unit, and each boarder should have its own equipment such as hand brush and sponge pads hanging inside its house. During the cleaning process the cat can be confined in its run or in a suitable carrier to prevent it paddling about in wet disinfectant. The attendant should wash his hands after handling or grooming each individual cat and in some catteries the attendant passes his waterproof boots through a pan of disinfectant solution before entering each run. Though such details might appear rather fussy and possibly unnecessary, they do stop cross-infection occurring and in the long term may save a great deal of expense and worry by preventing disease.

Indoor catteries are easier to clean than outdoor ones as they are not affected by wind and weather. Each boarder should have individual equipment, however, and the cats should not be allowed to exercise in a communal area.

All soiled litter and sweepings from the cat houses and runs must be disposed of carefully. The best method will be determined by the location of the cattery and by local regulations. Most catteries find it best to put all waste into sturdy plastic sacks to await disposal and whenever possible these should be incinerated. The sack should be carefully sealed to prevent contamination by flies and other insects which could cause the spread of infection.

Whenever possible catteries should be cleaned with an industrial vacuum cleaner. Cats are often frightened by the noise of such cleaners and must be safely confined in carriers while the machine is in use. The vacuum method enables every crevice of cat houses and runs to be cleaned of dirt, dust, shed hairs, fluff, fallen leaves, cobwebs and the eggs and larvae of parasites. After the walls have been vacuumed, floors and runs can be washed and allowed to dry thoroughly before the cat is freed. Window panes can be kept bright and sparkling with the use of a wash leather (window cleaning sprays may be toxic to cats).

In a breeding cattery all routines have to be flexible in order to fit in with the fluctuating feline population. At various times of the year there will be kittens at different ages and stages of development, newborn and nursing litters, weanlings and kittens ready for sale. Although all kittens must be kept in very clean accommodation, it is absolutely vital that the disinfectant used in cleaning their quarters is non-toxic and that all flooring and surfaces are thoroughly rinsed and dried before the kittens are allowed access.

At the critical nest-leaving stage of four to six weeks of age, small kittens explore all surfaces by licking, sniffing and tasting. Even non-toxic products, such as mild detergents, left in small areas may cause painful ulceration of the kittens' lips, mouths or tongues and can also give rise to serious gastric disturbances. Feeding dishes of small kittens should be carefully washed after each meal and should be of china or a plastic material suitable for sterilization with boiling water; otherwise disposable dishes should be used. Nest-leaving litters should be kept under observation to make sure that they do not lick and ingest the litter in their toilet tray. Some kittens go through a difficult stage when they may pass loose motions, due to teething. During this period it may be better to line their run with paper, change their toilet tray at frequent intervals during the day and wash the young animals' paws and tail ends whenever necessary.

In the cattery it should be a general rule to clean and attend those most at risk first, such as newly born and nursing kittens and queens approaching their birthing date. Unweaned and unvaccinated kittens should be dealt with next, then the growing and teenage stock, then barren queens and finally the stud cats. Any cats or kittens which have been exposed to infection by visiting another cattery or having taken part in a cat show, should always be left until last.

Whether it is used for breeding, boarding or both, a cattery will only prosper by following regular programmes of vaccinations, blood testing for diseases, good feeding and expert management.

8

Cat Care in the Home

CHOOSING YOUR KITTEN

Before deciding to buy a kitten, various factors must be considered. You must make up your mind whether you want to own a pedigree or non-pedigree cat, and whether you will want to breed kittens or just enjoy the companionship of a loving, neutered pet. You must also decide whether or not you want to embark on a show career, and become involved in the absorbing world of cat clubs and the Cat Fancy. You must examine your reasons for wanting a cat, and decide whether or not you are at home for sufficiently long periods to give an animal the necessary care and company. If you go out to work, you could consider having two cats, bringing them up together so that they become life-long friends and companions.

Consider carefully the various types of pedigree cats and only buy a longhaired variety if you are certain of keeping your resolution to undertake the very regular and thorough grooming sessions which are necessary. Think about the life you lead, and your temperament, and make sure that the type or breed that you choose fits in with both your lifestyle and your character. Some kittens are quiet and subdued in temperament, and some are very extrovert. Although breeds tend to have certain personality characteristics, there are of course in-dividualists, even in the same litter. If you have a growing family, you must choose a lively, bombproof type that has been reared in a busy environment, while cattery-reared kittens often do better in quieter, more peaceful homes.

Pedigree kittens should always be bought directly from their breeder and never through an agency or pet shop. Good, caring breeders want to meet prospective owners of their precious, carefully-reared youngsters, to be assured that each kitten will go to a suitable home, and be properly fed and well treated.

Before making up your mind about breeds, try to visit a large cat show. These are held in most major cities each year and are generally well advertised, though a list of shows may be obtained from the governing bodies or the feline press. Exhibitors at shows are generally pleased to discuss their own varieties with visitors, and by talking to several exhibitors you should gain a balanced impression of the various cats, their temperaments and habits. The official show catalogue lists each exhibit against its corresponding show pen number, and gives details of the exhibitor, the exhibit's name, age and sex, its parents and breeder. An alphabetical list of exhibi-tors is found at the back of the catalogue, and many breeders advertise too.

Many of the kittens on show will be for sale and most look very appealing, but do not be tempted to buy on impulse before exploring all the possibilities and being certain of the breed you really want. Kittens can be booked at a show, but resist the temptation to take one home with you. Shows often exhaust young kittens and lowered resistance in-creases the risk of infection. A slightly debilitated kitten is better off in its own home with its littermates, and so you should pay a deposit towards the cost of your chosen kitten and arrange to collect it two or three weeks after the show.

Kittens without pedigrees are obtainable from friends and neighbours, or from animal sanctuaries. You may also buy such a kitten in a pet shop or store, but then you must take extra care to ensure that it is properly weaned and has not contracted any infection from other kittens. Non-pedigree kittens are often advertised in newspapers and cat magazines, and you can then go to choose from the litter, and have the opportunity to see the mother cat.

Whether you decide on a pedigree or a pet, you should look for the same basic points in selecting your kitten. Firstly, the kitten must be properly weaned and should have been completely separated from its mother for at least two weeks before going to a new home. Many people mistakenly believe that a kitten that tucks in to three or four small meals each day may be classed as fully weaned. In fact it might still be taking a surprising amount of milk from its mother too, and its digestion could well be adjusted to this milk food. If such a kitten is suddenly taken away to a new environment and a new diet, it could suffer a severe setback and gastritis.

The kitten you choose should look clean and tidy, and have a firm slender body. The spine and hips should not protrude and there should be no sign of the pot-belly which could indicate the presence of parasitic worms. The eyes should be bright and clear, and the haw or third eyelid should be tucked away in the inner corner of each eye. The ears should be clean inside, with no dark scale which indicates canker. The mouth should look healthy, with pink gums and sharp white teeth and the nostrils should be clean and free from any sign of discharge. The coat should be soft and smell fresh. Part it along the spine near the tail and also behind the ears. If there are any tiny black grit-like grains here, it means that the kitten has fleas. Look under the tail too, for yellow staining in this region shows that the kitten is suffering from diarrhoea. If the kitten passes your examination, it is safe to take it home with you. If it fails, ask the breeder to treat it and arrange to collect it later.

If you decide to buy a pet kitten, you will probably be expected to take it at eight to ten weeks of age, and you must be careful to find out exactly how it has been fed, keeping to the same diet until it settles in your home. Keep it safely indoors too, away from infection, until it is old enough to have its course of preventive vaccinations.

A pedigree kitten will be older before it is allowed to leave the breeder's home, and depending on its age, it may also have completed its vaccination programme. When you purchase a pedigree kitten, you should receive a sheaf of paperwork too. Any conscientious breeder will supply a diet sheet and probably a list of helpful hints too. You will get a large form showing the kitten's pedigree extending back for three, four or even five generations. In Britain, a pink form, signed by the breeder, enables you officially to transfer the kitten's ownership into your name, but due to the time lag in registrations, the breeder may have to contact you later in order to give you the kitten's official registration number, for the completion of this form. If the kitten has had vaccinations, certificates completed and signed by the veterinary surgeon will also be passed over to you. In the USA, all the different registering bodies have their own systems.

TAKING A KITTEN HOME

When you go to collect your new kitten, be sure to take along a safe carrier, for it is very unwise to carry a kitten unconfined, either in your car or on public transport. Leaving home might prove a very traumatic experience for the young animal, and its behaviour may be unpredictable.

Carriers are available in a wide range of styles and designs, and constructed of various different materials. Top-opening carriers are best, for it is easier to extract an unwilling or frightened cat from above than to try to pull it forward out of a front opening. Do not be tempted to buy a small, kitten-size carrier, for your pet will soon outgrow this. Your carrier will last for many years and will prove a valuable investment.

Cardboard carriers are only suitable for emergency use, and should never be left unattended when your cat is inside. A determined cat is easily able to bite its way through cardboard, and if your cat wets the carrier, the bottom may fall out. The cardboard carrier is ideal, however, for use when a cat is suffering from a contagious disease, and must be transported to the veterinary surgery, after which the carrier and its bedding can be destroyed by burning, reducing the risk of spreading infection. Perspex and fibreglass carriers are very smart, and last for years. Their main drawback is the tendency to produce interior condensation. Some cats dislike being as visible as they are under the clear hoods of the perspex type, and seem to feel vulnerable. Wickerwork carriers are strong and sturdy. Cats love them and they can be covered with

Dual purpose carrier/sleeping basket

plastic or brown paper to keep out draughts. Their main drawback is that they are difficult to clean efficiently once soiled. Plastic-covered mesh carriers have become very popular in recent years and are strong, safe and very easy to wash and sterilize. Again, these are easily covered with plastic, polythene or strong brown paper, in order to be draught-proof and to give the cat an added feeling of security.

The carrier should be lined with soft paper towelling, topped with an old sweater or small piece of blanket, and if it is a cold day, a rubber hot-water bottle half filled with hand-hot water can be tucked between the layers.

On your return journey with your newly acquired kitten, resist all temptation to open the carrier's lid, no matter how vocally he protests at his confinement. Wait until you are safely within closed doors before you let him out. Fuss your kitten a great deal when you arrive home, and confine him to one smallish room at first, until he gets his bearings and gains confidence. The room should have been checked for safety before collecting the kitten. There must be no trailing flexes for him to chew with the possible danger of electrocution, no unguarded fire, no open window, no uncovered fishtanks, no impregnated insect killing strips, perfumed or household fresheners and no houseplants which, if chewed, could prove poisonous.

At first your new kitten will miss its littermates and so needs lots of extra kindness, comfort and play sessions. A cosy warm bed must be provided from the outset and as well as offering a suitable sleeping spot, should also act as a refuge whenever the little animal wants to rest, or to be alone. At first a small but substantial cardboard box makes an ideal bed. It can be disposed of when soiled, and can be replaced with boxes of gradually increasing size as the kitten grows. Kittens like the warmth and security of small boxes, and the sleeping box should be just large enough to allow it to stretch out across the base. Put a layer of newspaper in the bottom of the box for insulation, cover this with a thick pad of torn paper towels, then put a small piece of blanket on top for comfort. As the kitten turns into a mature cat, such boxes may be replaced by a smart sleeping basket or bed, complete with a proper cat cushion.

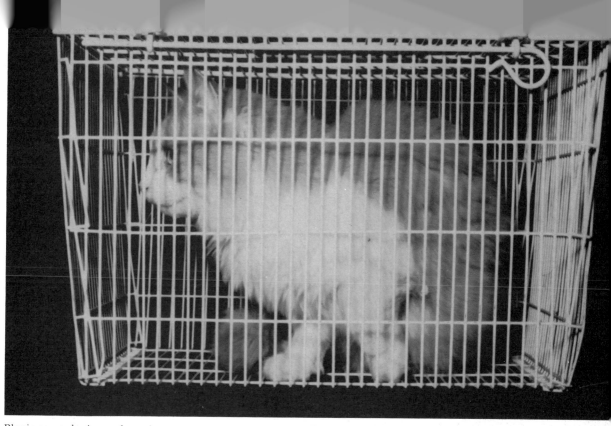

Plastic-covered wire mesh carrier

Isolation in a portable pen

General Kitten-care Timetable

Age of kitten	Treatment required
6 weeks	First dose of worming medicine
6–8 weeks	Complete weaning
8 weeks	Second worming
10 weeks	Third worming
12 weeks	Complete vaccination

Toys

Small cardboard boxes make good play houses for kittens. Holes of various sizes can be cut in all sides, and if some toys are placed inside, the kitten will have endless fun diving through the holes and inventing games. Children like to cover such boxes with washable wallpaper and show great inventiveness in constructing small houses and play castles for kittens.

There are many toys and games made expressly for cats, available in a wide price range. Some are well-made, safe and quite suitable as playthings for small kittens and some are potentially dangerous. Those to avoid include anything made of soft or pliable rubber or plastics, pieces of which may be chewed off and swallowed with dire consequences. Anything with stuck-on eyes or whiskers or adorned with tiny bells is rather risky. Simple toys are most acceptable to kittens and most felines love toy mice stuffed with catnip, and spiders made from pipe cleaners, bound together with wool to form the body, and with the wire ends safely bent over. Table tennis balls, feathers, cotton reels and small balls of crumpled paper all make ideal toys, encouraging play which improves co-ordination and muscle tone.

Toilet Trays

Small kittens needs shallow toilet pans, which should be kept clean and fresh at all times. Various types of litter may be used in the pan and at first the new kitten should have the same material as provided in its breeder's home. Kittens, especially longhaired varieties, are often trained to use shredded paper or torn paper towels. Wood shavings or peat moss are often used for shorthairs, and the

Cat flap

various proprietary brands of litter are excellent. Garden soil and ashes are rather messy, and have no deodorizing properties.

Solid wastes should be extracted from the pan as soon as the kitten has used it, and the pan's entire contents should be disposed of at regular intervals. The pan should be washed thoroughly, rinsed, then sterilized with a suitable, non-toxic disinfectant made expressly for feline use, or with a mild solution of bleach. If made of a suitable material such as polypropylene, it can be sterilized by rinsing with boiling water.

Toilet Training

When you first acquire your kitten and bring him home, have the toilet pan ready and set in its regular place on a sheet of newspaper. The little animal will want to use the pan shortly after eating and when it is seen to search around the edges of the floor, should be gently placed on the litter. Most kittens, having once used their pan, will always return to the same place, so be careful not to change the site of the toilet facilities for the first few weeks. When the kitten successfully uses the toilet pan for the first time, make a fuss of him, but if he should make a mistake and use the wrong place, ignore the fact, and clean up carefully, leaving no trace of any odour. Kittens are naturally very clean, and so long as their toilet pan is kept clean and fresh, they will rarely, if ever, use any other spot.

Toilet tray

'Superloo'

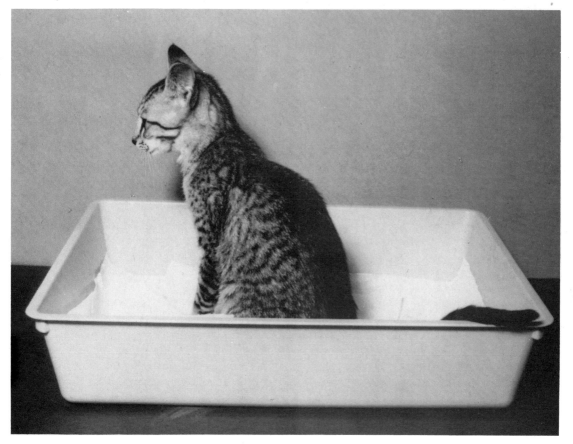

Claw Stropping

All cats need to strop their claws to remove scale, and if you do not supply a suitable scratching post, your pet might start to use the furniture. Specially constructed scratch posts can be purchased from pet shops. Some are made of plain wood, some are covered with carpet and some are of corrugated cardboard, impregnated with catnip.

Home-made scratch posts must be constructed so that the cat or kitten is able to stretch up to strop, and must be fixed firmly to a wall or onto a strong base so that they cannot topple over. Upright posts are ideal, and a log strapped to a leg of the kitchen table is perfect. Covering the post with carpet may not be such a good idea, for the kitten may think that any carpet surface is made to be stropped. Teach your kitten to use his post by gentle scolding when he strops at a chair or rug. Say 'No' sharply and clap your hands, then place him by his post and encourage him to place his paws on the rough surface.

FEEDING THE YOUNG KITTEN

When you first get your kitten it will need four small meals each day, fed at regularly spaced intervals. Some breeders wean their kittens onto all-meat diets, while others feed equal amounts of meat and milky feeds. This is why the diet sheet is so important and should be followed carefully for the first few weeks after the kitten's arrival in your home. Any upset to a small kitten's digestion can result in a devastating gastritis with diarrhoea and vomiting, followed by rapid dehydration and even death. A kitten used to having milk and cereal in its diet will normally have such a meal first thing in the morning and again late in the afternoon, and will enjoy a meat meal at midday and again in the evening. It is usual to allow a four hour gap between feeding meat and milk meals. The milk can be the evaporated variety, diluted with an equal quantity of hot water, and with enough pre-digested cereal stirred in to make a creamy paste. The cereals prepared for human babies are generally accepted by kittens, and those made from flaked rice seem to cause less stomach upsets than the oat and wheat derivatives.

Meat meals may be of raw, cooked or canned meat. Raw meat should be very fresh and chopped into very thin slivers rather than minced. It should be fed at room temperature and never directly from the refrigerator. Cooked meat may be of virtually any variety, and cooked in any way, but remember that some cooking methods destroy a considerable amount of its nutritional value. Again, feed at room temperature and make sure that there are no bones or skin on the kitten's plate. Canned food should be of the very best quality and preferably of a brand specially formulated for feeding to kittens.

Always give food on clean dishes, and make sure that these have been carefully washed, then thoroughly rinsed free from all traces of detergent, for this can prove toxic to kittens. Make sure that your kitten has access to fresh, clean drinking water at all times. The diet may be supplemented with vitamins and minerals if you wish, although a good varied diet should contain everything your kitten needs to remain healthy. If he cannot assimilate milk he may be slightly deficient in the calcium necessary to build a sturdy skeleton and strong teeth, so you may decide to add special calcium tablets to his diet until he stops growing. There are special complete diets for cats, dried cat chow and semi-moist pellet foods. All have been very carefully formulated to contain everything a cat needs for good health, and most owners make such products part of their pet's varied daily menu. The hard, complete dried foods are manufactured in a range of attractive flavours and special shapes which encourage cats to crunch them, helping to keep their teeth and gums healthy. Cats which do eat dried diets must be encouraged to drink extra water, and care must be taken not to overfeed with this concentrated fare.

FEEDING THE ADULT CAT

All cats need a diet which is high in protein and contains the right balance of essential vitamins and minerals. All lean meats may be fed to cats either raw or carefully cooked and are high in nutritional value. Raw liver acts as a mild aperient and is ideal for constipated cats; cooked liver is apt to have the reverse effect. Many top breeders feed lightly cooked liver to their cats once a week to provide sufficient vitamin A. Fish may be given, but again once a week is generally enough. Some cats like eggs and most enjoy milk, though not all cats are able to digest it. Cats are able to cope with a high fat content in their diet and butter or corn oil occasionally added to meals soon gives an added sheen to the coat. The better quality canned foods provide excellent well-balanced diets for all cats. Such food is prepared under strictly hygienic conditions, and is the result of years of research into feline needs.

The complete, dried diets are good provided that cats fed on them have unlimited access to fresh cool drinking water. Cats need very little carbohydrate and therefore it is unnecessary to add cereals or vegetables to their meals. Water, available at all times, should be freshly drawn each day and offered in a clean bowl.

Though vitamins and minerals are essential for the cat's well-being, a varied, well-structured diet will automatically provide these. Various vitamin supplements are available for deficient cats and those recovering from disease.

Calorie Requirements

The adult cat requires 200–300 calories per day, depending on its size, pattern of activity and amount of exercise. Kittens need a higher propor-

tion of calories in relation to body weight, due to their rapid rate of growth and their higher energy requirements.

Types of food	Approximate calorie value
Canned cat food	200 per small can
Dried diet	1,600 per lb
Lean minced beef	750 per lb
Liver	600 per lb
Heart	550 per lb
Melts (spleen)	450 per lb
Lights (lungs)	500 per lb
Rabbit	600 per lb
Chicken	500 per lb
White fish	500 per lb
Oily fish	750 per lb
Milk	400 per pint

NEUTERING

Unless your cat is to be used for breeding it should be neutered and will then make a perfect pet. Even if you wish to show your pet you may choose a neuter, for unlike those of the dog world, members of the Cat Fancy actively encourage neutering and there are special classes at all shows for de-sexed cats. Entire male cats are known to be very affectionate,

A Feeding Guide

Age	Average Body Weight	Daily food requirements	Number of feeds per day
	(lb)	(oz)	
Newborn	0.25	1.0	10
5 weeks	1.0	3.0	6
10 weeks	2.0	5.0	5
20 weeks	4.5	6.0	4
30 weeks	6.5	7.0	3
Adult male or female	7–10	6–8.5	1–2
Pregnant female	7.5	8.5	2–3
Lactating female	5.5	14.0	4
Neuter	9–10	6.5–8	1–2

but once mature spend hours in search of receptive females, if allowed their freedom, getting into fights and suffering from torn ears, lacerations and painful bite abscesses. As he reaches maturity, the urine of the tom develops the special pungent odour that allows him to mark out his territory, and he uses this trademark effectively from the age of about one year. He may be seen turning his back on various landmarks, raising his tail and pedalling with his hind legs allowing him to spray a few drops of urine high and wide. Male cats may be neutered at any age, but the general consensus of opinion between breeders is that 7–9 months is the ideal. By this age the young male has developed his masculine phys-ique but not his masculine habits, and when he is castrated he retains his truly male appearance. The neutering operation is simple and is carried out under a general anaesthetic by a veterinary surgeon, who asks for food and drink to be withheld from the cat for several hours before it is taken to the surgery. Post-operative recovery is rapid, but it is important to watch for any sign of distress, excessive licking of the scrotum or fresh bleeding. Most male cats eat and play normally again on the day following the operation, and neutered male cats generally live far longer than their entire brothers.

The female cat is neutered by an operation known as spaying, which entails the removal of the ovaries and uterus. The operation may be carried out at any time from four months of age, but most breeders prefer the kitten to be a little older. It is safer and easier to spay a cat before the first oestrus period occurs, and certainly before the cat has had kittens. It is very unfair to allow a cat to become pregnant, then ask for her to be spayed, simul-taneously terminating the pregnancy, for then the operation becomes much more serious.

In normal circumstances, the spaying operation is quite straightforward. The cat is given a general anaesthetic and an area of her flank is shaved, and swabbed. A small incision is made and the operation carried out. Very few stitches are needed to suture the incision, and the cat returns home once she has completely recovered from the anaesthetic. She must be kept quiet and encouraged to rest for several hours, then may lead a normal life. The stitches are generally removed about one week after

the operation, when the queen should also have another veterinary health check to ensure that all is well. The shaved area soon grows new hair, but in Siamese and other Himalayan-patterned cats, the regrowth is often very dark. At the next moult however, the new hair should grow through the correct colour once more. Judges at shows rarely penalize a dark mark in the spaying area, as they understand how the Siamese genes affect coat colour.

GROOMING THE LONGHAIRED CAT

All longhaired cats require regular daily grooming from kittenhood with a special session once a week when the ears, teeth and claws are checked and cleaned, if necessary. Cats with long, profuse coats of especially soft hair may need powdering every day to prevent matting. Those with coarser coats will probably need powdering only once a week. Specially formulated grooming powders are available or unscented baby powder may be used. The coat is brushed lightly from tail to head, parting the hairs, and the powder is sprinkled lightly into the roots, before being worked in gently with the fingertips. The powder must be left in the coat for about ten minutes while it absorbs grease and during this time a wire rake or special mitten may be used gently to tease out any knots or tangles. Take care that the wire teeth do not scratch the skin. When all the tangles have been removed, brush all the powder from the coat, working methodically from the tail end towards the head, then use a coarse blunt metal comb to ensure that every hair is separate.

When longhaired cats are neglected their coat become densely matted; the most serious knots form behind the ears, between the legs and under the tail and generally prove impossible to tease out. When this happens, the coat must be clipped or de-matted and if the cat is nervous or difficult to handle, this procedure has to be carried out under a general anaesthetic by a veterinarian. If a cat has to be clipped the new hair must receive daily attention as it grows and it is a good idea to keep the hair under the body clipped quite short, unless, of

GROOMING A LONGHAIRED CAT:
1 Apply powder

3 Tease out tangles

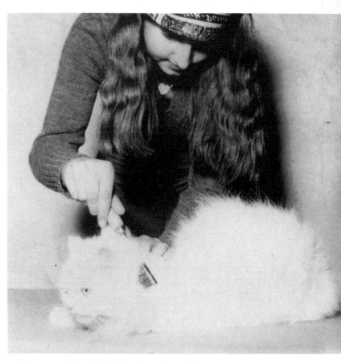

2 Rub powder into the coat to separate the hairs

4 Use a metal comb right through to the roots

course, the cat is destined for a show career. The inside of the cat's ears should be checked regularly. A dark exudate could indicate the presence of earmites and veterinary attention is required. Normal wax may be gently removed with a dry cotton bud, taking care not to probe into the delicate structure of the ear.

GROOMING THE SHORTHAIRED CAT

Most shorthaired cats need only routine grooming once a week, unless they are moulting heavily, or are being prepared for a show. If the coat is shedding, a rubber brush may be used very lightly to take out the loose hairs, otherwise a moderately

GROOMING A SHORTHAIRED CAT:
1 Clean the coat with a bristle brush

stiff bristle brush should be used from head to tail, lifting the coat and stimulating the circulation. Particular attention should be paid to the soft hair behind the ears, the throat, inside of each leg and under the tail and the coat should be checked for the black grits which indicate the presence of fleas.

A specially formulated coat dressing may be sprayed and worked into the coat, then a fine-toothed comb is used all over the cat, working from head to tail to remove all dead hair, scurf and dust. A chamois leather or a pad of warmed cotton wool may be used with a firm action over all the muscular areas of the body. This has a toning effect on the muscles and helps to raise a sheen on the coat. For very smooth-coated cats firm, gentle, hand grooming may be used instead of the polishing pad. Finally the eyes, ears and claw beds are examined and may be cleaned if necessary.

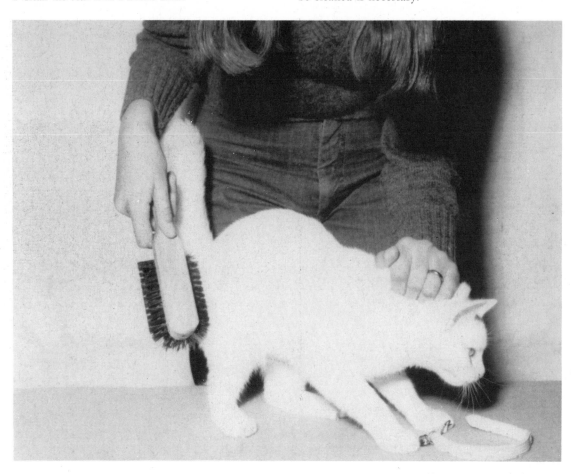

2 Spray on pest powder or coat dressing

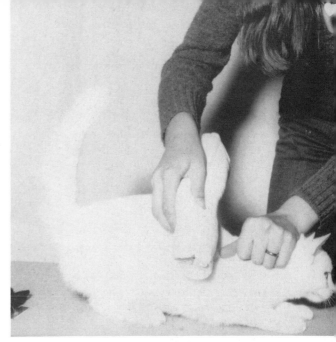

4 Buff the coat with a soft cloth

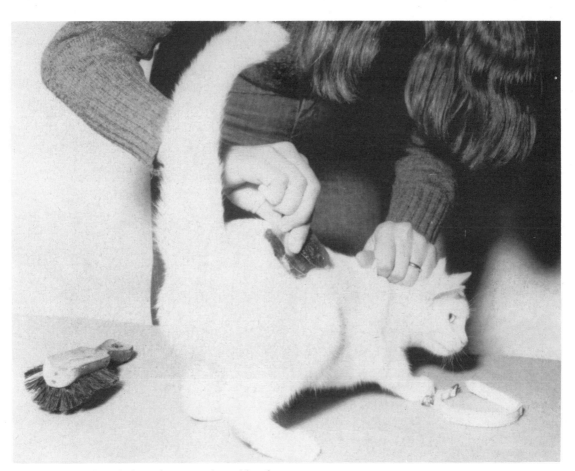

3 Remove dust, loose hairs and any parasites with a fine-toothed comb

GENERAL CARE FOR LONGHAIRED AND SHORTHAIRED CATS

Regular checks should be carried out on all cats, pedigree and non-pedigree. Once a week you should look at their teeth for signs of tartar build-up. If you notice any encrustations, take your cat to the vet to have its teeth scaled. While checking the teeth you should also check the gums for any signs of inflammation, for while this may be caused by eating sharp bones, it may also indicate the onset of an infectious disease.

With today's often unsatisfactory feline diets, many cats fail to get the necessary exercise for their teeth and jaws, and when mouth conditions are neglected the cat may be reluctant to eat sufficient food to stay healthy. All cats should have a lump of raw meat from time to time which acts better than any toothbrush.

Check the tail for signs of 'Stud Tail' – if this is present, you will see a greasy patch or little blackheads in this region. Wash the area with soap or mild detergent, rinse thoroughly, dry carefully and apply a little talcum powder, working it well into the skin. Although this condition is called 'Stud Tail', it is not confined to working male cats and may be seen in queens and in neuters. A similar condition is seen around the chin of some cats and if neglected, small abscesses may form.

Cleaning the ears

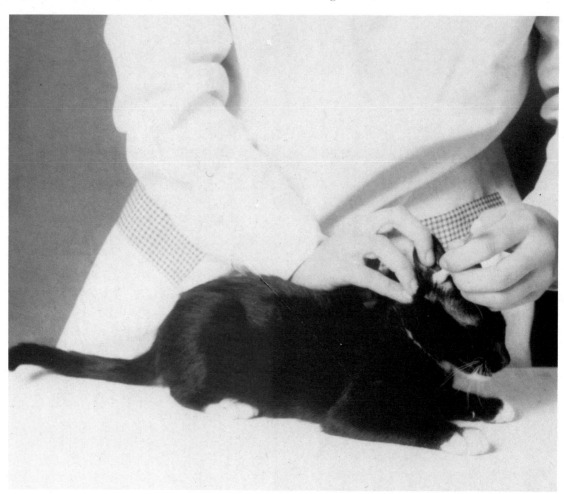

BATHING A CAT

If your cat becomes very soiled a wet bath may be given. Some cats enjoy the process while others hate it, so it is advisable to have an assistant at hand. A double sink is best; if this is not available you can use two large washing up bowls or baby baths. First carefully groom the coat, removing all knots, tangles and loose hair. Use water that is comfortably hot, testing the temperature with your bare elbow, just as you would check the water for a human baby's bath. Quarter fill one bowl and half fill the other, and have ready a good supply of warm towels, a shampoo made expressly for cats, mixed with warm water, and a container for ladling the water over the cat.

Stand the cat in the quarter-filled bowl, and wet its coat right through. Some feline coats repel water, so you may have to add a little shampoo to enable the water molecules to penetrate the pelt. When the coat is wet through pour shampoo into your palm, rub your hands together to form a lather, then massage in the shampoo, taking care to avoid getting water or shampoo into the eyes, mouth or ears. Pay particular attention to the underside of the cat, under the tail and the paws, particularly between the toes. If the cat struggles, or tries to climb out of the bowl, your assistant should take hold of the scruff so that you have both hands free to work up an effective lather. Lift the cat while the dirty water is emptied away, then replace the wet animal in the empty bowl and carefully rinse away every trace of shampoo by ladling clean water from the spare bowl. You may need to change the water occasionally for it is vital to ensure that every bit of foam is removed from the cat's coat. If you are able to use a sink or bath with a spray attached to the taps, so much the better, but be sure that the water remains at a constant temperature. After this thorough rinsing, use both hands to squeeze as much water as possible from the coat, then wrap the cat in a thick warm towel and rub as dry as possible.

Take care to choose shampoo specially formulated for cats, and not for dogs, and do not allow your cat to lick its coat until every trace of shampoo has been rinsed away.

Some cats allow themselves to be dried with a hand-held hair dryer; otherwise sit your pet in a mesh carrier next to a fan heater set to blow warm air. It is important to dry the cat as quickly as possible to prevent chills, and then the coat should be gently brushed and combed into place.

GIVING A BRAN BATH

A soiled or greasy coat may be cleaned by giving an old-fashioned bran bath. A double handful of bran is evenly spread over the base of a baking pan and put to heat in a moderately warm oven until it feels comfortably warm to the touch. Stand your cat on a table covered thickly with newspapers then take small amounts of the hot bran and massage it well into the cat's coat, starting at the root of the tail and progressing in stages towards the head. Work fairly quickly, using your fingertips gently and talking soothingly to the cat. When all the bran is used up, loosely wrap the cat in a large, warmed towel, and pet it on your lap. Most cats thoroughly enjoy this cossetting and warmth. After ten minutes, remove the towel and put the cat back on the newspaper while you methodically brush the bran from the coat. With it will come a suprising amount of dandruff, loose hair, dirt and dust. The next day groom the cat carefully to remove any last traces of bran, leaving the coat looking and smelling fresh and clean, with a healthy sheen.

DEALING WITH FLEAS

Fleas are small wingless insects with the capacity for jumping from one host to another, and from their hiding places in crevices in the cattery onto any warm body that passes. The insects suck the blood of cats, which can be very weakening, especially to kittens. They are carriers of disease and of the tapeworm, and their bite can set up an allergic reaction in some cats, so the sooner they are dealt with the better. Although the actual flea may not be seen on the cat, being nimble and very elusive, its presence is detected by the discovery of patches of black, gritty specks, the excrement of the flea, around the head of the cat, behind his ears, at the tail root and under the body. Fleas lay their eggs on the cat's fur and in crevices in the cat's bed.

Eggs laid on the fur soon drop off, spreading over a fairly wide area as the cat moves around. They fall into crevices and in the right warm and moist conditions will hatch out within ten days. From the egg comes a tiny worm-like larva which is not parasitic, but lives on minute particles of organic refuse, and grows rapidly, in due course spinning itself into a cocoon and pupating. The adult flea eventually emerges from the cocoon and waits for a warm-blooded animal to pass by onto which it will jump and start to suck blood.

Bearing in mind the life-cycle of the flea, it will be seen that the thorough cleansing of the accommodation, and the frequent changing of the bedding goes a very long way in the eradication of the flea, but it may be necessary to powder a cat which is very infested. A specially formulated powder or spray should be used which may be obtained from your veterinary surgeon. It must be carefully and methodically worked into the fur all over the cat, from the nose to the tip of the tail, under the body and down to the toes. Great care must be taken to avoid the eyes, the inside of the ears and the genital area. After powdering, pop the cat into an old, clean cotton pillowcase with a draw-string top. Pull up the string around the cat's neck and make a great fuss of him for ten minutes, then take him out, stand him on a sheet of newspaper and brush out the powder vigorously. Powder, dying fleas and debris will all fall onto the newspaper.

Shake out the pillowcase too – there will be dead fleas in there in all probability, and then roll up the newspaper and burn it. Shorthaired cats should be given two or three such treatments at weekly intervals, and longhaired cats five or six weekly treatments, after which the fleas should be completely eradicated.

The tapeworm has an interesting life-cycle. In the flea tapeworm the flea ingests the tapeworm egg and a minute embryo hatches, then burrows into the intestinal wall of the flea's body. It migrates freely within the flea, eventually becoming a bladder worm. A cat becomes infected by eating a flea which is host to a bladder worm and the worm then develops inside the cat in the form we know as the tapeworm. Rigid control of fleas prevents infection with tapeworm.

A-Z
of Feline Health

ABSCESS

Abscesses are often caused when skin heals rapidly over a deep cut, bite or wound, allowing bacteria introduced under the skin layers to multiply, forming a hot, painful lump. Abscesses caused by bites from other cats are commonly found on the head, paws or at the base of the tail. Before the lump becomes apparent, the affected cat may seem listless and, if checked, is found to have a raised temperature. Antibiotic treatment helps to disperse the swelling and relief can be given by hot fomentations. First clip the hair around the swelling, then apply hot, moist compresses for 20 minute periods several times daily until the abscess starts to drain. When the abscess opens, clean the area with 3 percent (10 volumes) Hydrogen Peroxide several times daily to prevent the formation of a scab. If the abscess continues to drain an offensive discharge for more than 48 hours, further veterinary treatment is essential.

ALLERGIES

Some cats are allergic to flea bites and develop a condition known as flea eczema. Small clusters of scabs form on the skin and the hair breaks away. Some plants cause allergic reactions in cats which exhibit symptoms such as conjunctivitis or asthma. Some cats, particularly those of Siamese-derived varieties, may be allergic to milk or other dairy products. Such allergies may manifest themselves in eczema-like conditions, gastric upsets or a combin-

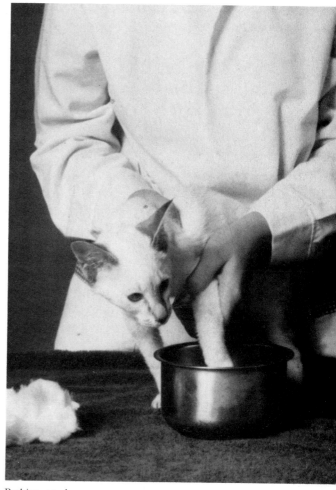

Bathing an abscess

ation of the two. Cats fed exclusively on fish may develop miliary or 'fish' eczema, when the skin along the side and flanks becomes dry and intensely itchy, causing the cat to scratch and lick the region, devoiding it of hair. Eggs cause a similar reaction in some cats and some drugs also produce allergic reactions. Avoidance of the cause of the allergy is the best method of treatment.

ANAESTHETICS

As cats are difficult to handle when in pain or in strange surroundings it is necessary to anaesthetize them for even minor operations. It is important to withhold food and drink for about 12 hours before the administration of any anaesthetic, to ensure that the stomach is empty, and to prevent choking. Recovery rates vary and usually depend upon the length of time that the cat has been unconscious. The cat should be kept quiet, warm and confined in a carrier or pen until completely conscious. Half-conscious cats may stagger and fall, and as their reflexes are not normal, they often react to being touched and may bite.

ANAL GLANDS

Situated under the tail, the tiny anal glands secrete a sebaceous fluid, designed to impart a subtle scent to the animal's faeces. Occasionally the tiny openings in the glands become blocked and small abscesses form. The affected cat is seen to drag its seat along the ground and spend much time licking the anal region. Professional treatment is necessary to clear the impacted or infected glands.

ANTIBIOTICS

Antibiotics are drugs used extensively for the treatment of bacterial infections. If antibiotic treatment is prescribed for your cat you should be careful to give the correct dosage over the prescribed time period. It is important that the antibiotic level is maintained in the cat's body by administering small, regular doses at the proper times. Antibiotic treatment must be given only under veterinary supervision (it is important to administer the full course) and any unused pills or medicine should be discarded and never kept for future use.

ANTISEPTICS

Many of the antiseptic products manufactured for use in humans are toxic to cats. The safest antiseptic for feline use is 3 percent (10 volume) Hydrogen Peroxide, with a final rinse of salt water.

ARTIFICIAL RESPIRATION

There are several ways of reviving an unconscious cat. Gentle swinging by the hind legs, the head hanging downwards, may prove to be effective. Alternatively, lay the cat in a natural position on its side, press its chest gently, then relax. Repeat rhythmically, every four seconds, and while there is a recognizable heart beat there is still hope of recovery.

Mouth to mouth resuscitation is generally effective. To do this, turn the cat on its side, hold its mouth closed and place your own mouth over its nose. Blow steadily into its nostrils for three seconds, then take a deep breath and repeat until some resistance is felt or the chest is seen to rise. After one minute stop and check the chest to see if the cat is breathing on its own. If it is not, continue as above until veterinary assistance is at hand.

ASPIRIN

Never give aspirin to your cat; it can be fatal, as can many medicines manufactured for human use.

BLADDER INFECTION

A cat that starts to urinate outside its litter pan or is seen to squat for long periods of time without urinating could have a serious bladder infection. Other signs of this disease include blood in the urine and excessive licking of the genital region. A cat with any of the symptoms above should be taken to the veterinary surgeon without delay.

BLEEDING

Any wound which spurts blood indicates that an artery has been cut and professional treatment is necessary, but until professional help is available a clean folded handkerchief should be pressed firmly over the bleeding point. If the bleeding is from a limb, a temporary tourniquet may be applied and this is done by tying a handkerchief around the limb above the wound and twisting a pencil beneath it until the material is just tight enough to stop the blood flow. Release the tourniquet after ten minutes, otherwise irrevocable tissue damage will occur. A cat haemorrhaging from any natural opening must be kept warm, still and quiet until veterinary help is at hand. Slight bleeding from minor wounds may be treated with antiseptic (see above).

BONES

Small bones may get stuck in the cat's mouth or throat, causing such panic and distress that the cat paws frantically at its mouth and drools streams of saliva. Wrap the cat firmly in a towel and have an assistant hold him while you check the mouth using a spatula or the handle of a teaspoon to hook out the bone. If you cannot find the cause of the problem or you are unable to remove a bone, confine the cat in a carrier and rush it to the vet.

BRONCHITIS

Bronchitis may develop after a respiratory disease and very careful nursing is needed. The veterinary surgeon will adminster antiobiotic treatment but you can play your part in your pet's recovery by helping to keep its breathing passages open. Inhalations are very beneficial. (See Home Nursing and Inhalation).

CALCIUM

This mineral is an essential part of the domestic cat's diet. Wild cats obtain their calcium by eating the whole carcase of their prey, but the cosseted house cat may need added calcium, especially during the early months of its life, to ensure proper bone formation and healthy teeth. Pregnant and lactating queens need a higher level of the mineral in order to produce strong, healthy kittens. Milk provides adequate amounts of calcium but if cats cannot digest it a special form of sterilized bone meal or veterinary calcium tablets should be added to the diet.

CANKER

Canker is a term used to describe trouble in the cat's ear. It is usually caused by tiny mites which live and breed deep down in the ear canal and can only be completely eradicated by specialist treatment. A special liquid is generally instilled into the affected ear at regular intervals, and the treatment should be carried out under veterinary supervision. Proprietary brands of so-called 'canker powder' should never be put into the cat's ears. Such products quickly dry out the normal healthy moist condition of the ear, and are unlikely to kill earmites, but may form into a solid, caked mass and produce serious ear infection. Cats dislike liquids being dropped into their ears, but it the dose is drawn into a plastic syringe which is allowed to stand in hot water for a few moments the liquid assumes blood heat and the cat is less likely to object to the treatment. Shake the syringe well and test its temperature against the inside of your wrist before administering. Put three or four drops in the ears then massage behind the cheek bone to work the liquid right down to the site of the trouble.

CLAWS

The retractile claws of the cat may tear or break off causing swelling and lameness. Torn claws often bleed profusely but pressure of your thumb over the bleeding point should stop the haemorrhage. Swelling of the claw bed may be relieved by soaking the foot in a solution of Epsom Salts (magnesium sulphate heptahydrate). Mix a dessertspoon of Epsom Salts in half a pint of boiling water and stir until dissolved. Reduce the temperature until it is comfortably hand hot, and pour into a clean empty jam or coffee jar. Sit the cat comfortably on your lap on a folded towel and hold the affected limb inside the jar for five to ten minutes. Repeat the treatment as often as is practicable during the day and following instructions as for Abscess (page 123).

CONJUNCTIVITIS

This may be caused by an allergy, an irritant in the eye or herald the onset of an upper respiratory disease. Conjunctivitis is an inflamed condition of the eye's outer membranes and its treatment consists of bathing the eye with saline solution using fresh swabs for each application. If the condition does not rapidly improve veterinary treatment is required.

CONSTIPATION

This condition can be caused by any of several factors. Longhaired cats may become constipated after swallowing a large amount of loose hair through self grooming. Normally active cats may be affected when confined in a boarding cattery and other cats may become constipated due to a drastic change of diet. Having made certain that your cat is in fact constipated and is not straining due to a bladder infection (see above), home treatment consists of giving copious fluids and small meals of oily fish, such as sardines or pilchards. Should the condition continue, seek veterinary advice.

DEHYDRATION

This serious condition needs immediate veterinary treatment. It is one symptom of the most serious of cat diseases and may be the prime cause of death in some. The dehydrated cat suddenly appears thin and drawn and if the skin of the scruff is pulled up and away from the neck, it remains lifted and feels dry and hard, whereas normal skin springs back into place. No time should be wasted in getting the cat to the vet.

DIARRHOEA

This condition is most often caused in young kittens by feeding a faulty diet. It may be caused by milk, eggs or offal and only experience will prove which foods or combinations of foods cause the production of frequent fluid and offensive motions. Diarrhoea may also be caused by intestinal worms or by toxic substances such as disinfectants, pesticides and some gardening products. No cat should be allowed to suffer from diarrhoea for more than three days or dehydration (above) may occur; a vet should be consulted.

EYE INJURIES

If your cat rubs or paws at his eyes or the eyes are red or discharging, there might be a foreign body in the eye or an injury might have been sustained. Emergency treatment consists of flushing the eye by pouring cool, boiled water into it from a plastic bottle. Never use boracic powder or eye-drops prescribed for humans. If the soreness persists, seek veterinary advice.

FLEAS

As fleas can transmit several diseases among cats and may also act as an intermediate host to the tapeworm it is important to keep your cat flea-free. A very fine-toothed metal comb is ideal for checking a cat's coat for it will lift flea excreta from the hair and occasionally detect the adult flea. Special sprays are manufactured for use on cats and are very effective though they may cause side effects when used on young kittens and pregnant or nursing queens. Fleas do not hatch in the cat's coat, so attention should be paid to the animal's bedding and resting places. Thorough vacuum cleaning should lift the minute eggs and larvae which live in house dust.

FRACTURES

Fractures may be sustained by cats involved in accidents, and immediate veterinary attention is essential before swelling occurs. Great care must be exercised in moving or restraining an injured cat until help is at hand. A seriously injured cat can be moved gently onto a flat tray or piece of wood which may then be used as a stretcher. Unless you are skilled do not try to apply splints for you may do more harm than good and cause unnecessary pain to your pet.

GINGIVITIS

This is a condition in which the gums become sore and inflamed, showing a bright red line at the tooth margins. Veterinary treatment is essential as gingivitis could be the first sign of a serious illness developing in your pet.

HAW

This is the name given to the nictitating membrane which forms the third eyelid of the cat. In a sick cat the haw may show like a thick triangular skin at the inner corner of the eye. Although this is not a specific symptom of disease, the visible haw indicates that your cat either has a temperature, is carrying a heavy burden of parasitic worms or is feeling generally unwell.

HOME NURSING

The caring cat owner institutes a regime of regular veterinary checks and vaccinations. Your veterinary surgeon can be one of your cat's best friends and his advice should be sought and carefully followed whenever any health problems occur. It must be remembered, however, that although skilled veterinary treatment can be administered at the surgery, home nursing plays an important part in the cat's recovery from any serious or prolonged illness.

The three essentials of good nursing are: maintaining the cat's will to live, providing adequate liquids and nourishing foods and keeping the patient clean and warm.

Each feline patient and each illness has characteristics which make it unique. A sick cat must be carefully confined in a quiet area shielded from light, and away from other pets. Heat must be provided by spot heating the patient's bed with an infra-red lamp. The nursing area must be easily cleaned and all equipment should be suitable for sterilization by heat or strong disinfectants. Try to use disposable equipment whenever possible. Perhaps the most demanding disease in terms of time, care and patience, is Cat 'Flu, for a seriously affected cat may be so depressed and miserable that it resents care and handling and is quite prepared to die. In acute cases, ulceration of the mouth causes severe

Syringe feeding an orphan kitten

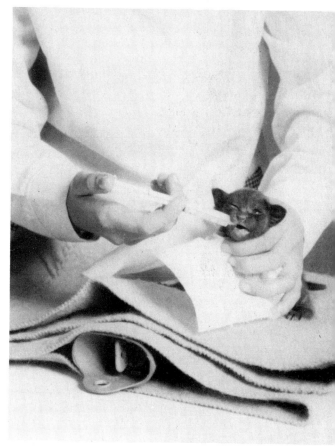

pain and even gentle syringe feeding with liquids may be difficult.

The nostrils, lips and eyelids may also be ulcerated and could bleed when bathed, and severe diarrhoea scalds the anal region which also bleeds. The vet will give skilled antibiotic treatment but such severely sick cats also need very special home nursing. Concentrated liquid foods are readily available, and these may be drawn into plastic syringes for feeding. The nozzle of the syringe is passed between the teeth at the side of the animal's closed mouth, then the plunger should be gently depressed, allowing a few drops of the liquid to be swallowed. Take care to prevent choking and allow the cat to take a breath between swallows. Remember, its nose may be so full of discharge that it must breathe through its mouth.

After feeding the cat should be cleaned. First bathe its eyes, nose and mouth, using separate swabs for each area, dipped in warm saline solution. Dry the areas carefully and soothe any ulcerated regions with an application of petroleum jelly. If the cat is well enough, place it on its toilet tray and stroke it encouragingly. Bathe the anal region if soiled and apply petroleum jelly if it is ulcerated, then dust with talcum powder, as used for human babies. Work the powder well into any soiled areas of the coat, massage gently with the finger tips, then brush out again.

Incontinent cats should have disposable napkins or diapers, as designed for human babies, torn into convenient shapes and wrapped neatly round the hind region. These should be changed frequently in order to keep the animal very clean.

Depressed or dehydrated cats benefit from having their paws and legs massaged. First the feet should be gently rubbed, then the hind legs should be stretched and flexed.

A cat that is unable to stand must be changed onto a different side frequently to prevent the development of sores on the pressure points.

Each day, try to tempt the cat to eat voluntarily by offering strong smelling foods such as kipper, herring, crab or lobster available in tiny pots as paste and spend as much time as possible with the sick cat, providing various forms of stimuli in an attempt to instil the will to live.

INHALATIONS

To give an inhalation treatment, first smear petroleum jelly around the eyes and nostrils of the cat. Put it in a mesh carrier, without any lining or blankets. Make up the inhalation according to the manufacturers' instructions with boiling water and put in a small bowl in the centre of a large empty washing-up bowl, placed on the floor. Place the carrier squarely on top of the washing-up bowl and cover with a plastic or folded cotton sheet, to contain the vapours. After ten minutes, release the cat who will be greatly relieved with streams of mucus coming from its previously blocked nasal passages. Wipe all the discharges away and repeat the treatment as often as is practicable during the day.

INJECTIONS

The veterinary surgeon usually uses the injection method to ensure that your cat receives the correct doses of antibiotics or other substances necessary for its treatment or welfare. Many vaccinations are given by subcutaneous injection which is quite painless. Injections may be given intramuscularly or intravenously, but these are slightly more difficult to administer and the cat often needs some restraint. If you find injections upsetting, ask a friend to take your pet to the vet, for your distress may affect your cat.

JAUNDICE

If your cat's eyes change colour and the mucous membranes around the mouth and nose take on a yellowish hue, it may have jaundice, a single symptom of several major cat diseases, so immediate professional advice should be sought.

KINKS

Small deformities in the tail bones of cats are called kinks. They cause no harm but as the fault is hereditary, cats with kinks should not be used for breeding.

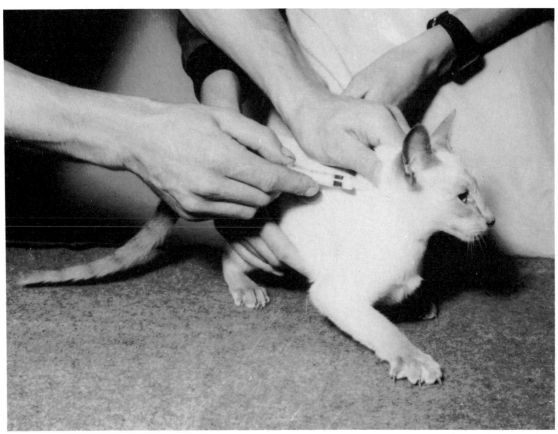

Giving a subcutaneous injection

LICE

A cat infested with lice scratches vigorously at its head and neck. The tiny, slow-moving parasites may be seen in the coat and the bead-like eggs are cemented to individual hairs. The veterinary surgeon will provide the necessary sprays and shampoos for their eradication. All combings should be burned and the cat confined until the last louse is destroyed.

LICKING

Excessive licking by the cat of any area of its body may herald the onset of eczema or could indicate the presence of some foreign body, such as a grass seed, a splinter or a tick. First try to determine the cause and treat it if possible, but if there is no obvious explanation for the cat's behaviour, seek professional advice.

MANGE

This unpleasant skin condition is caused by tiny mites which burrow beneath the skin of the cat. The animal scratches, looks generally unwell and may develop small bald patches, usually around the head. Mange is highly contagious and can be difficult to cure, so veterinary treatment is essential, and your cat must be isolated until it is completely cured.

POT-BELLY

A kitten with a distended stomach is said to have a pot-belly and this condition usually indicates a heavy infestation with parasitic worms or points to a faulty diet. Obtaining expert diagnosis is very important, followed by routine treatment and care.

PULSE

The normal pulse rate of an adult cat is between 110 and 120 beats per minute. The pulse is best felt by placing the middle finger inside the cat's thigh along the course of the femoral artery.

STOMATITIS

Inflammation of the mouth may be due to any of several causes, but an affected cat stops eating and soon loses weight. The cause may be very hot food or sharp bones, but true stomatitis may be the first stage in the course of a serious infection, so seek expert advice.

TEMPERATURE

The normal temperature of an adult cat varies between 38°C and 38.6°C (100.5°–101.5°F). While a slight rise in temperature might be caused by any number of reasons such as stress or fear, a sub-normal temperature should always be regarded as a danger sign. Temperature taking is resented by some cats and so an assistant is necessary, to take a gentle but firm hold of the animal's scruff. A veterinary thermometer with a blunt end should be liberally lubricated and inserted into the rectum for a few moments.

TICKS

Free-ranging cats occasionally pick up ticks. These are first noticed during grooming as bluish-coloured, wart-like growths attached to the skin.

Taking a kitten's temperature

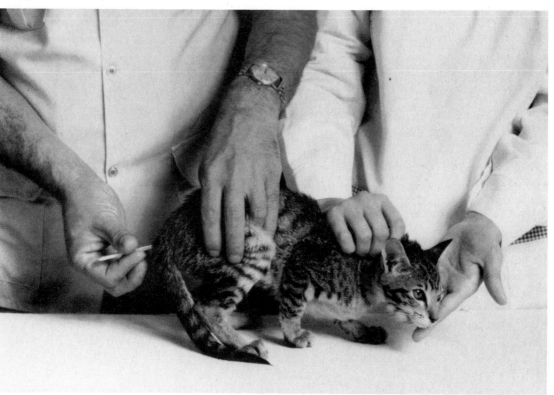

Ticks suck blood until fully engorged, then drop off the cat, but in the process they are very debilitating. A drop of surgical spirit applied to the point of attachment to the cat's skin causes the tick to relax its jaws and it may then be removed with tweezers. Never try to pull the tick away or you may leave the jaw behind, when an abscess will form in the cat's skin.

VIRUSES

Four of the many infectious feline diseases are of viral origin and are considered to be of major clinical importance.

Feline Panleucopenia (F.P.L.) or Feline Infectious Enteritis (F.I.E.)

Highly contagious with an incubation period of two to nine days, this disease is characterized by its very sudden onset. The cat shows profound depression, refuses all food and may vomit yellow bile or white froth. Its temperature may be as high as 40.5°C (105°F) in the early stage of F.P.L. then rapidly reduces until it is subnormal. The process of dehydration is so rapid that the cat appears visibly to shrink and sits hunched up with its coat erect and staring, its eyes glazed. It may hang its head over its bowl, though it will not drink. If touched or lifted the cat cries in a despairing manner and its body feels cold and rigid to the touch. Death is normally rapid but if the cat does live for more than a few days, blood-stained diarrhoea is generally seen. A cat with F.P.L. needs careful nursing and intensive veterinary care. It is best confined within a mesh carrier or kitten pen, and must be strictly isolated. It is important that everything used in nursing is disposable. Metal pens can be finally sterilized by intense heat or by being immersed in a solution of formalin. Dishes, syringes and bedding should all be burnable, the nurse should wear easily sterilized clothing and avoid contact with all other cats during the course of the illness and quarantine period. The mortality rate is very high and animals that do recover from F.P.L. may shed virus for some time, so should be isolated. Recovery is long and very slow. The usual period of quarantine for catteries which have suffered an outbreak of F.P.I. is six calendar months from the final recovery date.

PREVENTION

Effective vaccines are available for protecting cats against F.P.L. Preliminary injections are given to kittens and booster doses may be administered annually or bi-annually.

Feline Leukaemia Virus (FeLV)

FeLV was isolated in 1964 and was found to thrive in the lymphoid tissue and bone marrow of the cat. These areas of the body produce blood cells and the main diseases caused by this virus are forms of leukaemia and anaemia. Leukaemia is a malignant disease of the white blood cells and an affected cat generally shows a very gradual weight loss. Various treatments have been tried for feline leukaemia, without success. The virus is known to be transmitted by direct contact between cats or passed vertically from a pregnant queen to her kittens. It is thought that cats are able to build up natural immunity to FeLV if they are subjected to minute doses of the virus at widely spaced intervals. Research continues into the dread disease and its incubation period seems to vary from a few weeks to several years.

PREVENTION

As this virus is most usually transmitted by direct contact, a degree of control is possible. Caring cat breeders follow fairly rigid systems to prevent Fe.L.V. infection invading their stock. All their cats are regularly blood tested and any showing a positive result are removed from the cattery. Later, repeat sampling is carried out and if all the cats show negative results, the stock can be considered free of Fe.L.V. All newcomers or visitors to the cattery are isolated and tested, and with such care the disease should be kept at bay until such time as an effective vaccine is available.
Note: Fe.L.V. has been discovered in the treatment of other diseases, therefore it seems feasible to suspect the presence of this virus in any cats that become progressively thin or listless, are infertile, or abort their kittens.

Feline Viral Rhinotracheitis (F.V.R.) and Feline Calicivirus (F.C.V.)

These two viruses are often found together in the cat and give rise to the syndrome known as Cat 'Flu. F.V.R. is the most common and severe of feline respiratory diseases; it is highly infectious and has an incubation period of two to ten days. Its first symptoms include general listlessness, complete loss of appetite and sneezing. As the cat's temperature rises, eyes and nostrils begin to discharge and long streams of saliva may hang from the mouth. Secondary infections often occur, producing conjunctivitis and in old cats and young kittens, broncho-pneumonia may develop.

There are a number of strains of Calicivirus, compared with the single strain of F.V.R. and therefore they produce a wide range of symptoms from a severe upper respiratory infection, as seen with F.V.R., to a mild sub-clinical infection. Ulceration of the mouth is typical in F.C.V. and may be the only sign of illness in the cat. Whether the cat is infected with F.V.R., F.C.V. or both viruses, veterinary treatment should be sought without delay. Antibiotics are given to prevent and cure secondary bacterial infections and vitamin injections may be administered to counteract depression and loss of appetite. A cat with Cat 'Flu should be isolated from other cats and must be carefully nursed throughout its illness. Daily Vitamin C given in combination with anti-biotic treatment has been found to aid recovery and small meals of tempting food are essential.

PREVENTION

Vaccines are available to immunize cats against respiratory diseases and should be used intelligently to set up and maintain Cat 'Flu-free catteries. Healthy kittens from the age of 12 weeks may have the vaccine at the same time as the injection against F.P.L. Some types are administered intra-nasally, others by injection, and your veterinary surgeon will advise on the best vaccination programme for your cat.

VOMITING

Cats naturally regurgitate their food, especially when they have eaten too quickly or too well. When vomiting produces yellow fluid or white foam and is accompanied by diarrhoea or a high temperature it should be treated as a symptom of something serious. If your cat vomits for more than four hours or has the other symptoms too, contact your veterinary surgeon.

WORMS

Intestinal worms are present in many cats. They are unpleasant parasites, very debilitating and should be eradicated. As there are several types of worms and each needs different treatment, it is advisable to collect a small sample of the cat's faeces for examination by a veterinary surgeon. Microscopic analysis soon helps to identify the parasite and the correct medicines may be prescribed. It is not advisable to give your cat proprietary brands of worming pills which may be harmful, causing gastric upsets.

YEAST TABLETS

Your cat's daily requirement of Vitamin B may be given in the form of yeast tablets. Some cats will eat such tablets as treats and an overdose can do no harm. It is a fallacy, however, to believe that a daily dose of yeast tablets will act as a panacea for all feline ills.

ZOONOSES

Diseases transmitted between cats and humans are happily quite rare and to avoid them you should follow some simple rules. Keep your cat free from parasites; cook his food and wash his dishes separately from your own; do not let him lick your face, do not let him sleep in your bed; and wash your hands carefully after touching him or his toilet tray.

Appendices

COLOUR DISTRIBUTION OF PEDIGREE CATS

Self-Colours

Genetic Colour/Pattern	Longhair/Persian	Shorthair	Oriental/Foreign	Others
White	Copper-eyed (U.S.A.) Orange-eyed (U.K.) Blue-eyed Odd-eyed	Orange-eyed Blue-eyed Odd-eyed	White	
Black	Black	Black	Black	Sable Burmese (U.S.A.) Brown Burmese (U.K.) Bombay (U.S.A.)
Blue	Blue	Blue	Foreign Blue Russian Blue Korat	Blue Burmese
Chocolate	Chocolate Self	Chocolate Self	Oriental Chestnut (U.S.A.) Havana (U.K.) Havana Brown (U.S.A.)	Chocolate Burmese
'Light' Chocolate			Cinnamon	
Lilac	Lilac	Lilac	Foreign Lilac (U.K.) Oriental Lavender (U.S.A.)	Lilac Burmese
'Light' Lilac	—	—	Light Lilac (U.K.) Caramel (U.S.A.)	—
Red	Red	Red	Red	Red Burmese
Cream	Cream	Cream	Cream	Cream

Tabby (Agouti)

Genetic Colour/Pattern	Longhair/Persian	Shorthair	Oriental/Foreign	Others
Black	Brown	Brown	Brown	Sable Egyptian Mau (U.S.A.) Usual or Ruddy Abyssinian
Blue	Blue	Blue (U.S.A.)	Blue	Blue Abyssinian
Chocolate	—	—	Chocolate	Bronze Egyptian Mau (U.S.A.) Chocolate Abyssinian
'Light' Chocolate	—	—	Cinnamon	Sorrel Abyssinian
Lilac	—	—	Lilac	Lilac Abyssinian
Red	Red	Red	Red	—
Cream	Cream (U.S.A)	Cream	Cream	Cream Abyssinian

Parti-Colour

Genetic Colour/Pattern	Longhair/Persian	Shorthair	Oriental/Foreign	Others
Tortoiseshell:	Tortoiseshell	Tortoiseshell	Tortoiseshell	Tortoiseshell Burmese
Blue	Blue-Cream	Blue-Cream	Blue-Cream	Blue-Cream Burmese
Chocolate	Chocolate-Tortie	—	Chocolate-Tortie	Chocolate-Tortie Burmese
Lilac	Lilac-Tortie	—	Lilac-Tortie	Lilac-Tortie Burmese
Bi-colour:	Bi-color (U.S.A.)	Bi-color (U.S.A.)	—	—
(Any colour + white)	Bi-colour (U.K.)	Bi-colour (U.K.)	—	—
Tri-colour:	Calico (U.S.A.)	Calico (U.S.A.)	—	—
Black/red/white	Tortoiseshell and White (U.K.)	Tortoiseshell and White	—	—
	Dilute Calico (U.S.A.)	Dilute Calico (U.S.A.)	—	—
Blue/cream/white	Blue Tortie and White	Blue Tortie and White	—	—

Silver

Genetic Colour/Pattern	Longhair/Persian	Shorthair	Oriental/Foreign	Others
Tipped:				
Black	Chinchilla	Tipped (black) (U.K.) Chinchilla (U.S.A.)	Tipped	—
Red/cream	Shell Cameo	Tipped (U.K.) Shell Cameo (U.S.A.)	—	—
Shaded:				
Black	Shaded Silver Pewter (U.K.)	Shaded Silver (U.S.A.)	—	—
Red/cream	Shaded Cameo	Shaded Cameo (U.S.A.)	—	—
Smoke:				
Black	Smoke	Smoke	Smoke all colours	—
Blue	Blue Smoke	Blue Smoke	—	—
Red/cream	Cameo Smoke	Cameo Smoke (U.S.A.)	—	—
Ticked:				
Black	—	—	—	Silver Abyssinian
Spotted	—	Silver Spotted	Silver Spotted	—
Tabby	Silver Tabby	Silver Tabby	Silver Tabby	—

Note: Rex cats are found in most feline coat colours and patterns, as are Manx, and in the United States the Scottish Fold, the Turkish Angora, the Maine Coon and the Exotic.

Himalayan

Genetic Colour/Pattern	Longhair/Persian Himalayan (U.S.A.) Colourpoint (U.K.)	Shorthair Colourpoint (U.K.)	Oriental/Foreign Siamese	Others
Black	Seal-point	Seal-point (U.K.)	Seal-point	—
Blue	Blue-point	Blue-point (U.K.)	Blue-point	—
Chocolate	Chocolate-point	Chocolate-point (U.K.)	Chocolate-point	—
Lilac	Lilac-point	Lilac-point (U.K.)	Lilac-point	—
Red	Red-point	—	Red Colorpoint SH (U.S.A.)	—
			Red-point Siamese (U.K.)	—
Cream	Cream-point	—	Cream-point Siamese (U.K.)	—
Tabby	Tabby-point (all colours)	Tabby-point (all colours)	Lynx Color Point (U.S.A.)	All colours
			Tabby-point Siamese (U.K.) (all colours)	All colours
Tortie	Tortie-point (all colours)	—	Tortie Color Point (U.S.A.)	All colours
			Tortie-point Siamese (U.K.)	All colours

Note: Birman cats are Himalayan in pattern with the addition of white paws – they are found in serveral points colours. Balinese cats are longhaired Siamese and are found in several points colours.

138

2. GESTATION TABLE

January

MATED	1 2 3 4 5 6 7 8 9 10 11 12 13 14 15 16 17 18 19 20 21 22 23 24 25 26 27 28 29 30 31
KITTENS	7 8 9 10 11 12 13 14 15 16 17 18 19 20 21 22 23 24 25 26 27 28 29 30 31 1 2 3 4 5 6
	MARCH APRIL

February

MATED	1 2 3 4 5 6 7 8 9 10 11 12 13 14 15 16 17 18 19 20 21 22 23 24 25 26 27 28
KITTENS	7 8 9 10 11 12 13 14 15 16 17 18 19 20 21 22 23 24 25 26 27 28 29 30 1 2 3 4
	APRIL MAY

March

MATED	1 2 3 4 5 6 7 8 9 10 11 12 13 14 15 16 17 18 19 20 21 22 23 24 25 26 27 28 29 30 31
KITTENS	5 6 7 8 9 10 11 12 13 14 15 16 17 18 19 20 21 22 23 24 25 26 27 28 29 30 31 1 2 3 4
	MAY JUNE

April

MATED	1 2 3 4 5 6 7 8 9 10 11 12 13 14 15 16 17 18 19 20 21 22 23 24 25 26 27 28 29 30
KITTENS	5 6 7 8 9 10 11 12 13 14 15 16 17 18 19 20 21 22 23 24 25 26 27 28 29 30 1 2 3 4
	JUNE JULY

May

MATED	1 2 3 4 5 6 7 8 9 10 11 12 13 14 15 16 17 18 19 20 21 22 23 24 25 26 27 28 29 30 31
KITTENS	5 6 7 8 9 10 11 12 13 14 15 16 17 18 19 20 21 22 23 24 25 26 27 28 29 30 31 1 2 3 4
	JULY AUGUST

June

MATED	1 2 3 4 5 6 7 8 9 10 11 12 13 14 15 16 17 18 19 20 21 22 23 24 25 26 27 28 29 30
KITTENS	5 6 7 8 9 10 11 12 13 14 15 16 17 18 19 20 21 22 23 24 25 26 27 28 29 30 31 1 2 3
	AUGUST SEPTEMBER

July

MATED   1 2 3 4 5 6 7 8 9 10 11 12 13 14 15 16 17 18 19 20 21 22 23 24 25 26 27 28 29 30 31

KITTENS 4 5 6 7 8 9 10 11 12 13 14 15 16 17 18 19 20 21 22 23 24 25 26 27 28 29 30 1 2 3

SEPTEMBER   OCTOBER

August

MATED   1 2 3 4 5 6 7 8 9 10 11 12 13 14 15 16 17 18 19 20 21 22 23 24 25 26 27 28 29 30 31

KITTENS 5 6 7 8 9 10 11 12 13 14 15 16 17 18 19 20 21 22 23 24 25 26 27 28 29 30 31 1 2 3 4

OCTOBER   NOVEMBER

September

MATED   1 2 3 4 5 6 7 8 9 10 11 12 13 14 15 16 17 18 19 20 21 22 23 24 25 26 27 28 29 30

KITTENS 5 6 7 8 9 10 11 12 13 14 15 16 17 18 19 20 21 22 23 24 25 26 27 28 29 30 1 2 3 4

NOVEMBER   DECEMBER

October

MATED   1 2 3 4 5 6 7 8 9 10 11 12 13 14 15 16 17 18 19 20 21 22 23 24 25 26 27 28 29 30 31

KITTENS 5 6 7 8 9 10 11 12 13 14 15 16 17 18 19 20 21 22 23 24 25 26 27 28 29 30 31 1 2 3 4

DECEMBER   JANUARY

November

MATED   1 2 3 4 5 6 7 8 9 10 11 12 13 14 15 16 17 18 19 20 21 22 23 24 25 26 27 28 29 30

KITTENS 5 6 7 8 9 10 11 12 13 14 15 16 17 18 19 20 21 22 23 24 25 26 27 28 29 30 31 1 2 3

JANUARY   FEBRUARY

December

MATED   1 2 3 4 5 6 7 8 9 10 11 12 13 14 15 16 17 18 19 20 21 22 23 24 25 26 27 28 29 30 31

KITTENS 4 5 6 7 8 9 10 11 12 13 14 15 16 17 18 19 20 21 22 23 24 25 26 27 28 1 2 3 4 5 6

FEBRUARY   MARCH

Index

Numerals in *italic* type indicate the main reference to the entry; those in **bold** type indicate pages on which illustrations appear